AF706947

The Ethics of Artificial Intelligence

Digital Sciences Set
coordinated by
Abdelkhalak El Hami

Volume 5

The Ethics of Artificial Intelligence

Edited by

Franck Debos

iSTE

WILEY

First published 2025 in Great Britain and the United States by ISTE Ltd and John Wiley & Sons, Inc.

Apart from any fair dealing for the purposes of research or private study, or criticism or review, as permitted under the Copyright, Designs and Patents Act 1988, this publication may only be reproduced, stored or transmitted, in any form or by any means, with the prior permission in writing of the publishers, or in the case of reprographic reproduction in accordance with the terms and licenses issued by the CLA. Enquiries concerning reproduction outside these terms should be sent to the publishers at the undermentioned address:

ISTE Ltd
27-37 St George's Road
London SW19 4EU
UK

www.iste.co.uk

John Wiley & Sons, Inc.
111 River Street
Hoboken, NJ 07030
USA

www.wiley.com

© ISTE Ltd 2025
The rights of Franck Debos to be identified as the author of this work have been asserted by him in accordance with the Copyright, Designs and Patents Act 1988.

Any opinions, findings, and conclusions or recommendations expressed in this material are those of the author(s), contributor(s) or editor(s) and do not necessarily reflect the views of ISTE Group.

Library of Congress Control Number: 2025939936

British Library Cataloguing-in-Publication Data
A CIP record for this book is available from the British Library
ISBN 978-1-83669-015-3

The manufacturer's authorized representative according to the EU General Product Safety Regulation is Wiley-VCH GmbH, Boschstr. 12, 69469 Weinheim, Germany, e-mail: Product_Safety@wiley.com.

Contents

Author Presentation

Horea Mihai Badau

Horea Mihai Badau is an ATER at the University of Lorraine, a researcher at the Centre de recherche sur les médiations (CREM) and a qualified lecturer in the 71st section. He is a member of the *Société française des sciences de l'information et de la communication (SFSIC)*. Author of two books on the digital world, his research interests include the ethics of artificial intelligence, digital communication in organizations and online journalism.

André Boyer

André Boyer is Professor Emeritus of Management Sciences at the Université de Nice Sophia-Antipolis. He has taught marketing and management in Nice and abroad, and directed numerous theses. He has published several books on organizational strategy and management, notably on corporate ethics, as well as a large volume of articles, including a recent series on AI and organizational management. He has organized several management training programs abroad, notably in Morocco, the Czech Republic, Madagascar, Indonesia and China. He recently headed the IPAG campus in Nice. Previously, he headed the IECS Strasbourg and chaired the public management jury at ENA.

Alexandre Carbonneau

With a doctorate in human sciences (information and communication sciences), Alexandre is currently a researcher at the SicLab Méditerranée laboratory at the Université de Nice Sophia-Antipolis. He teaches Human Sciences and Communication at EFREI Paris. His work focuses on the consequences of technoscientific developments (nanotechnologies, genetic engineering, etc.) on individual and social existence. Alexandre Carbonneau also tackles the issues surrounding the implantation of artificial intelligence in our brains, as well as the effects on our idea of ourselves, in the sphere of our relationship with technical objects.

Juan Antonio Carrasco

Juan Antonio Carrasco is a sociologist with a master's degree in philosophy from the University of Chile and a doctorate in sociology from the Autonomous University of Barcelona. He is an active member of the GEPS ("Globalization, Education and Social Policy") research laboratory. His main lines of research include various themes of political philosophy, with a focus on the critical implications of technological change, as well as the study of processes of socio-educational transformation. Mr. Carrasco stands out for his analysis of social and educational dynamics in the context of globalization.

Franck Debos

Franck Debos is a lecturer in Information and Communication Sciences at the Université de Nice Côte d'Azur, and is interested in the issues associated with the ethics of AI. His fields of study and experimentation include smart cities and the resulting digital uses for citizens/consumers, and smart tourism in a regenerative organizational logic. Through his teaching duties at universities and business schools, as well as his status as a researcher at the Université Nice Côte d'Azur (PhD in Management Sciences and qualified in Information and Communication Sciences), he combines conceptual and operational thinking. He has authored several articles and parts of collective works in the above-mentioned fields.

Faranak Farzaneh

Associate Professor of Management and Artificial Intelligence at IPAG Business School. Faranak Farzaneh has studied, taught and conducted research in France and abroad, notably in Canada and the UK. She is responsible for the incubated Chair of AI and Ethics, as well as being the academic director of the European PEPAICO project. Her research, published in academic journals, focuses on people and AI, the internal dynamics of boards of directors and innovative employee behavior. She is attached to the IPAG LAB laboratory and the IPAG Chair in Inclusive Enterprise.

Maria Daniella Lancini

Maria Daniella Lancini teaches at the Université Côte d'Azur and in business schools. She holds a PhD in Information and Communication Sciences, a Graduate Course in Risk Communication from Harvard University and a Master 2 in Communication and Ecology and Sustainable Development from the Université Nice Côte d'Azur. Maria Daniella Lancini is a specialist in international cooperation (fundraising and management of high-level projects for UN agencies, the WHO, the World Bank, etc.). In her native Venezuela, she has held senior positions in public institutions such as the Ministries of the Environment and Health.

Vanessa Landaverde Kastberg

Vanessa Landaverde Kastberg is a research associate at the SIC.Lab Méditerranée laboratory at the Université Côte d'Azur, where she obtained her doctorate in information and communication sciences. Her research focuses on digital communication and social media. She teaches digital marketing and advertising at several business schools in France, China and Monaco. She also supervises professional theses in business administration. Thanks to her role as a consultant for consultancies in France and Monaco, she is able to monitor the evolution of digital communication and the impact of AI.

Matina Magkou

Matina Magkou is an associate researcher at the SIC.Lab Méditerranée laboratory (Université Côte d'Azur), teacher and consultant in cultural and creative industries and international cultural cooperation. Her research focuses on cultural cooperation and policies, intercultural communication, as well as on cultural and creative industries and the impact of AI at this level. She holds a PhD in Leisure Sciences, Communication and Culture from the University of Deusto in Spain. Between 2023 and 2024, she was a visiting researcher at the Centre for Digitalisation, Democracy and Innovation at the Brussels School of Governance (VUB). She is currently a Fulbright Schuman Scholar at New York University and co-founder of ΚΟΛΕΚΤΙVA for Social Innovation and Culture.

Nicolas Oliveri

Nicolas Oliveri is a teacher-researcher at IDRAC Business School. He holds a doctorate and a Habilitation to Direct Research (HDR) in Information and Communication Sciences, and is a member of the SIC.Lab Méditerranée laboratory at the Université Côte d'Azur in Nice. His work focuses on the technical imaginary, organizational communication, e-learning, technosocial risks and artificial intelligence. He currently heads the NUMERISC research program on the digitization of organizations, ICT ethics and responsible management.

Loïc Sumien

Loïc Sumien holds a doctorate in information and communication sciences and is a research associate at the SIC.Lab Méditerranée laboratory. His research focuses on public policies for media and information literacy. Specialized in public and political communication, he has successively held the positions of communications director for the town of Saint-Tropez and the Cannes Pays de Lérins agglomeration community.

Introduction

This book addresses the issue of the need for ethics to support developments in Artificial intelligence (AI), whatever the field concerned. AI is not only profoundly changing our practices, but also the way we think and interact, at the level of individuals, organizations and societies, in our private, public and professional lives.

However, as a preamble, we feel it is important to try and define the contours of this sector of activity, which is increasingly present in our daily and professional lives. At this level, we can observe the existence of numerous definitions of AI, making it difficult to precisely define its contours.

This multiplicity of approaches to AI can originally be explained by its multidisciplinary genesis. Indeed, AI's official birth certificate can be traced back to the Macy Conferences in New York between 1942 and 1953, which brought together mathematicians, logicians, anthropologists, psychologists and economists to study ways of reproducing human learning and intelligence through computer technology.

In this context of reflection on the definition of AI, we must also take into account the two main structural axes of this notion, from the 1950s onwards, namely:

– an axis centered on a symbolic approach that was prevalent until the 1980s, and which approached the notion of intelligence through the use and

Introduction written by Franck DEBOS.

manipulation of symbols based on formal rules (*Physical Symbol Systems* by Newell and Simon in 1976);

– a so-called “connectionist” axis, inspired by biological neural networks and based on the fact that perception or intelligent behavior is the consequence of interactions between various computational units managed by learning rules (see the pioneering work of McCulloch and Pitts as early as 1943, or, closer to home, that of Levesque in 2013, who defines AI as an approach to intelligent behavior based on computational logic).

This notion is the subject of much debate among specialists in the field (Lecun 2019), and is even regularly called into question, with the emphasis more on the recognition capabilities of the machine or device under study, rather than on intelligence itself.

Some recognized specialists in this field even consider the possibility that AI is not as intelligent as it is made out to be, and that it is not intelligence that characterizes today’s AI systems, but their ability to recognize through machine learning. Based on this approach, some authors prefer to speak of augmented intelligence (Carter and Nielsen 2017) or even go so far as to say that AI does not exist, insofar as the concept of intelligence is not precisely defined and the debate remains very open at this level (Julia 2019; Karppi and Granat 2019).

The literature proposes three families of definitions, each of which represents a complementary facet of what AI is and what it can achieve. The first family is based on a technical or technological understanding of AI, and presents it as a branch of the digital sciences (or computer science).

For some, AI is defined as a family of statistical models, i.e. predictive models based on data processing, using machine learning techniques. According to this definition, AI is a technique, as it describes a set of processes (algorithms) or methods capable of obtaining a given result. AI is also a technology in that it is a discipline for studying techniques specific to AI, and can therefore be understood as a field in which new techniques can be developed.

The limitation of exclusively technical or technological definitions is that they restrict the field of possibilities, and ignore functional evolutions that could lead to the emergence of new techniques. It forces us to look at AI from the small end of the telescope.

A second family defines AI as a machine capable of reproducing or simulating human intelligence (i.e. learning, understanding, reasoning, interacting, etc.). Therefore, a *science* of AI would be the study of ways to automate the processes of perception, learning, reasoning, decision and action.

The World Intellectual Property Organization (WIPO) supports this view when it defines AI as "a discipline of computer science that is aimed at developing machines and systems that can carry out tasks considered to require human intelligence, with limited or no human intervention".

However, highlighting the ability of AI to simulate or reproduce human deliberation, decision-making and action processes enables a wide audience to appropriate the concept and give free rein to the imagination of possible futures. These imagined futures can range from the most extravagant fantasies or unjustified fears, to potentially productive anticipations and the awareness needed to define the rules governing them.

A third family of definitions adopts a more functional perspective. It presents AI as a set of artificial systems capable of adapting more or less autonomously to its environment, to achieve predefined ends. The definition proposed by the OECD in its report on *Artificial Intelligence in Society* (2019) is along these lines: "An AI system is a machine-based system that can, for a given set of humandefined objectives, make predictions, recommendations, or decisions influencing real or virtual environments. AI systems are designed to operate with varying levels of autonomy."

This definition is inspired by the initial approach proposed by Simon in 1969 in his book *Sciences of the Artificial*, and emphasizes the ability to interact with an environment on which the AI system can act and thus contribute to modifying. This definition also makes it possible to envisage different classes of systems, depending on whether they are "closed" or "open" to a predefined end (the objective assigned to the machine). As a result, it is possible to envisage and study different classes of behavior, without being constrained by particular technical considerations.

This functionalist approach focuses on behavioral characteristics. It allows us to approach the study of AI systems not only as distinct, identifiable systems whose function is predefined, but also as a set of interacting subsystems whose overall behavior is not always predictable.

This question becomes particularly relevant with the multiplication of system forms, the explosion of data produced (notably due to the multiplication of connected objects) and immediately available, and the progressive delocalization of AI algorithms as close as possible to data-producing objects (via Edge Computing).

This latter conception could be closer to the method favored by the European Commission, which is based on a risk-based approach (April 21, 2021). Indeed, this recent proposal for a regulatory framework is based on a time-tested definition of AI, according to a risk-based approach (unacceptable, high, limited, minimal). Article 3 of the draft regulation defines an AI system as software that can, for a given set of human-defined objectives, generate results such as content, forecasts, recommendations or decisions influencing the environments with which they interact.

These three families of definitions complement each other and can be combined to address issues relating to the impact of AI on society, the economy and, in particular, ethics. It is also important to consider the structuring of scientific studies around AI. For example, the possibility of interactions between heterogeneous systems brings us back to issues close to the study of the behavior of biological systems, such as animals or animal societies. This is how a group of scientists has initiated work using experimental techniques inspired by ethology to analyze AI (see the article *Machine Behavior* published by *Nature* in 2019).

In this context, and integrating the performance achieved by today's learning machines, our approach and vision of AI must take into account the possible emergence of "conscious" AI, equivalent in the first instance to human AI (Goertzel 2014) and capable of surpassing it (Boström 2014). Our definitions of AI cannot exclude reflections on so-called "strong" AI as opposed to "weak" AI, corresponding to current developments in this notion[1]. This reflection allows us to approach AI from an angle other than

1 Some authors, such as Collins (2018), propose a more refined typology of AIs. He distinguishes six levels of AI: the first refers to current technologies, the second corresponds to the notion of "asymmetrical prostheses" that can perform a human activity (Google's Alexa or chatbots) without being able to spot and compensate for our mistakes. The next four levels are of the order of a certain forward-looking vision: 1) "symmetrical prostheses" that correct a person's errors; 2) machines and devices capable of totally or partially reproducing the internal processes of human intelligence; 3) autonomous anthropomorphic machines that can form communities or

that based on the reproduction of human intelligence (or other forms of intelligence present in certain animal species). In the context of so-called "strong" AI, a machine or device with an intelligence equal to or even superior to that of a human being could then, in order to achieve a goal, use all the means at its disposal without taking into account ethical notions and what Russell (2019) calls the "preference structure". The latter takes into account only those solutions that are compatible with the human being and that, for him, are implicit (provided, of course, that the intentions are good).

AI therefore represents one of the major societal challenges, whose implications and applications are playing an increasingly prominent role in our lifestyles, and in technological developments linked to the collection and analysis of massive data. Through the tools it develops, AI can considerably improve our future, but at the same time bring with it the development of numerous risks and challenges with potentially worrying gray areas, notably in terms of democratic life and the use of eminently sensitive personal data.

After years of completely ignoring these dimensions, and turning AI into a vast social, ethical and political experiment on which citizens, associations and the vast majority of companies had not been consulted, it has become clear that ethical problems are now at the heart of AI's crucial issues: bias, discrimination and human rights are just a few examples.

In fact, the design and use of these new technologies must be supported by serious ethical reflection. It is worth pointing out that, on an international level, many societal actors (citizens, NGOs, institutions, companies, etc.) confirm the urgent need to reflect ethically on AI, leading to the development of an ethics of (or for) AI. This reflection raises a number of questions, such as whether AI falls within the scope of general ethics or whether it is specific. If it is specific, are we not seeing the strong influence of a consequentialist ethic that is often overly utilitarian? Should we broaden our thinking on AI ethics to include other ethical approaches, such as those found in moral philosophy (deontology, autism and axiologism)?

Another important question we need to address today is how to define an ethics by design for AI, which is not limited to applying major theoretical currents or principles that have been thought out upstream, or to assessing

---

even societies; and 4) machines with a physicality different from ours, that can self-improve over generations and also live in society.

the impacts or consequences downstream. Ethics by design are situated at every stage of the AI design, development, dissemination and feedback process, and offer a reflexive dynamic between ethical theories and concrete or practical cases, allowing non-majoritarian points of view or positions to be made visible or audible, in addition to those commonly adopted and which may be otherwise contradictory.

The main aim of this book, divided into three parts, is to shed light on the issue of AI ethics from the perspective of different disciplinary fields and sectors of activity.

The first part has a more conceptual dimension and refers to the imaginary of AI. Here, Nicolas Oliveri approaches AI through the lens of Robotic Humanities. His starting point is the observation that the research issues surrounding Digital Humanities can be found in, and are developing around, a variety of scientific disciplines. On an epistemological level, this concept is still in a state of flux, and refers to the questioning of human future in the face of the technologization of the world. In this context, on an exploratory and intercultural level, "Robotic Humanities" present an imaginary future, within which robotic technology takes on a mostly salvific dimension (see above all American cinema, and Japanese manga or anime). In fact, Nicolas Oliveri presents an analytical deconstruction of the mechanisms behind this utopian vision of post-humanist thought, in which strong social representations play an essential role in the study of human–machine fusion with regard to AI, whether through "director–creators" or "spectator–receivers".

In this part, Juan Carrasco reflects on the frontiers of technology and the future of AI. His starting point is the increasing use of the IOT (Internet Of Things) by individuals in complex digital environments. The degree of precision and processing power of AI associated with human activities is leading to a certain questioning and evolution of the traditional boundaries of technology and communication. This mutation forces us to question the ethical implications of these uses, which is the climax of his chapter. The author develops an approach based on three key concepts to better understand the limits of technology and the ramifications of AI. The central point of his reflection is how Derrida's politics of the archive intersects with Marcuse's and Heidegger's exploration of the foundations of modern technology, in order to better grasp the ethical implications created by the strong evolution of generative AI in the field of communication.

Juan Antonio Carrasco thus encourages us to re-imagine our approaches to interactions between individuals and their coexistence with the multiplicity of "intelligent" objects that make up our environment, going beyond the rhetoric and even the "technologist" and "millenarian" invocations, with a broader view of the implications for our democratic system and societal structures.

To conclude this part, Alexandre Carbonneau discusses the impact of technoscientific developments on mental ecology. Indeed, companies such as GAFAM (or GAMMA since Facebook is now called Meta) or BATX in China, as well as governments, are investing in experiments paving the way for a homo-silicon with enhanced logical–mathematical skills through brain implants. His reflections focus on the impact of such neuro-augmentation on our perception of ourselves, others and our societies. In this chapter, Alexandre Carbonneau focuses on the new relationships and interactions between the individual and the technical and technoscientific ecosystem he has helped to create. He points out that, until the end of the last century, the innovations imagined and developed by human beings enabled them to extricate themselves and surpass their natural condition, while retaining control over the rationality of their technical system. Since then, however, and particularly in the course of this century, the ever-increasing pace of technological innovation (especially nanotechnology) has tended to invalidate this observation. This state of affairs is helping to create a form of irrational sanctification of the sciences and associated techniques, which could lead to a new societal order centered on the (increasingly realistic) myth of the "augmented" human being, endowed with intellectual and physical super-power.

The second part highlights the need for an ethical framework in the context of integrative uses. The two chapters that make it up illustrate this "ethical obligation", through the lens of two booming business sectors.

In this part, André Boyer and Faranak Farzaneh look at the impact of AI on love, and more specifically at the role of dating sites in relationships. These sites bear witness to the increasing "platformization" of our lives, in the same way as e-commerce. The authors consider this strong affective feeling as leading to specific behaviors, and it can be considered a key element in social relationships, as it integrates not only social norms and values, but also and above all people's feelings. As a result, as with most of our social interactions, which are increasingly evolving in hyper-connected spaces, our love relationships, according to the authors of this chapter, are both supported and questioned by the evolution of social networks associated with digital platforms, which encourage us to make a set of choices similar to consumer behavior, in order to identify the person in phase with our expectations and desires. According to the authors, we are witnessing a challenge to the "traditional" model of romantic love as the dominant paradigm for relationships. In this context, through online dating, we are witnessing a total erasure of the first "face-to-face" unexpected encounter, as well as the emotions associated with it. André Boyer and Faranak Farzaneh therefore analyze the impact of digital technology on love relationships and their perception, notably through social networks, dating sites and, indeed, the AI associated with these media, through the prism of a theory for love.

The second chapter in this part deals with the importance of AI in the design and development of smart cities, and the ethical considerations associated with it. Franck Debos reminds us that this vision of the city is based on the interconnection of networks by cross-referencing the data emitted by each of them, in order to set up a new urban ecosystem to respond to societal evolutions. The author points out that current changes in cities, whether organizational, technological or societal, must not only integrate the issues inherent in the ecological transition, but also be associated with social considerations. In this context, we need to rediscover the foundations of sustainable development within the global approach initiated by the smart city, in order to develop participative governance and experienced management of natural resources, all with a view to satisfying the needs and expectations of institutions, businesses, associations and citizens. In this article, the author outlines the major challenges and avenues of action for moving away from the highly techno-solutionist approach that dominates smart cities, towards an approach that enables these urban territories to have a positive environmental and social impact. In this context, the role of AI is becoming increasingly important, and we need to think about an ethical framework for the use of AI in smart city projects, in order to better understand the risks of abuses and to ensure that the human dimension is at the heart of this digitalized vision of the city.

The final chapter of this part follows on from the previous one in terms of reflections and field of study. In this chapter, Maria Daniella Lancini and Franck Debos take as their starting point the observation that, in the articles, books, white papers, etc., dealing with the success factors of a smart city, there is relatively little discussion of the importance of the role of companies at this level. The aim here is to identify the main issues and expectations of French companies with regard to smart cities, irrespective of their sector of activity or size. At the same time, the authors examine the evolution and perspective of these same companies once they have joined the smart city ecosystem. These different reflections will enable us to define the value of their impact on the transformation of a "conventional" city into a smart city. In this context where AI is omnipresent, the authors will also address the need for regulated AI to develop relationships between the actors considered.

The final part of this book addresses the ethics of AI at the level of professional communities that are undergoing major changes as a result of the digitization of their activities. At this level, Loïc Sumien addresses the impact of AI on media and information literacy. Indeed, AI, which has existed for decades in the media sector and is associated with digital social networks, is significantly restructuring people's modes of communication and information, and more broadly the information ecosystem through new production and distribution media accessible to all. The author also points out that the interest aroused by AI is very recent, and can be associated with the emergence of tools that have appeared in the last two years (ChatGPT, Midjourney, etc.), opening the way for reflection on the ethical uses and apprentissages of this technology. To illustrate the issues raised by AI, Loïc Sumien examines the media and information literacy sector, where new pedagogical practices associated with AI are developing in close connection with information and communication sciences. These practices highlight the need for ethical, experienced and thoughtful use, as well as the importance of AI, continuing to reinvent itself and meet the expectations of the audiences concerned.

At the same time, Horea Mihai Badau is looking at AI algorithms, unknowingly present in our smartphones, which act as active "mediators" in terms of digital communication, selecting and interpreting information in the same way as news editors. The author has created an AI algorithm simulator similar to TikTok's recommendation system, and developed a user interface to make interactions easier. Several principles of "algorithmic thinking" emerge, associated with user usage, optimization of given goals and dependence on historical data. Horea Mihai Badau then carried out an analysis of the nodes (in this case, influencer accounts) in line with the algorithms, as well as an analysis of the impact of AI algorithms on users' semantic universes, through the recurring patterns and themes existing in their interactions. This research has identified the motivations of individuals most impacted by algorithms, and highlights the need for an AI ethical framework at this level.

The final contribution in this part is presented by Vanessa Landaverde Kastberg and Matina Magkou and deals with the ethical challenges of AI use in the creative cultural industries (CCI), as well as the debates and responses they generate within international bodies, in this case UNESCO and the EU, starting from the conceptual foundations and starting points of these institutions in terms of commitments to ethical AI. The authors also reflect on the relationship between this technological field and the needs and rights of creative people within CCIs. Their approach is based on documentary research into the ethical issues surrounding AI, and the two institutions mentioned above. This work is enriched and completed by interviews with representatives of these organizations as well as interest groups at the CCI level, as well as by collective reflections emanating from international colloquia dealing with the relationship between AI and CCIs.

References

Boström, N. (2014). *Superintelligence: Paths, Dangers, Strategies*. Oxford University Press, Oxford.

Carter, S. and Nielsen, M. (2017). Using artificial intelligence to augment human intelligence. *Distill*.

Collins, H. (2018). *Artifictional Intelligence. Against Humanity's Surrender to Computers*. Polity Press, Cambridge.

Goertzel, B. (2014). Artificial general intelligence: Concept, state of the art, and future prospects. *Journal of Artificial General Intelligence*, 5(1), 1–46.

Julia, L. (2019). *L'intelligence artificielle n'existe pas*. First, Paris.

Karppi, T. and Granata, Y. (2019). Non-artificial non-intelligence: Amazon's Alexa and the frictions of AI. *AI & Society*, 34(4), 867–876.

Lecun, Y. (2019). *Quand la machine apprend. La révolution des neurones artificiels et de l'apprentissage profond*. Odile Jacob, Paris.

Levesque, H. (2013). On our best behaviour. *IJCAI*, Beijing.

McCulloch, W.S. and Pitts, W. (1943). A logical calculus of the ideas immanent in nervous activity. *Bulletin of Mathematical Biophysics*, 5, 115–133.

Newell, A. and Simon, H.A. (1976). Computer science as empirical inquiry: Symbols and search. *Communications of the ACM*, 19(3), 113–126.

Russell, S. (2019). *Human Compatible: Artificial Intelligence and the Problem of Control*. Viking, London.

PART 1

A Theoretical and Imaginary Approach to AI

1

Imaginary Fantasies of Robotic Humanities and AI in Fictional Contexts

1.1. Introduction

Like many other hopeful technologies, representations of artificial intelligence are inherently Manichean. From its advent in the 1950s, to its socio-cultural and politico-economic entrenchment in the 2010s, AI has generated as much excitement as it has fear. This exacerbated bipartisanship seems to reflect the discourses that have always accompanied the history of technology and its implicit corollary, namely the movement towards technical progress, which is supposedly inescapable. In this respect, the Digital Humanities have been sharpening expert debates in recent years, as they have become involved in particularly broad research issues, themselves nested at the heart of disparate scientific disciplines. In the epistemological race to stabilize a concept that is still in a state of flux and whose definition remains open to question, we need to raise the question of what it means to be human in the face of the technologization of the world. To this end, in an exploratory and cross-cultural framework, Robotic Humanities propose highlighting the imaginary dimension in the construction of a future in which robotic technology is perceived as a saving force (notably in American cinema, and Japanese manga or animation). The Information and Communication Sciences (ICS) do indeed question the fictional and imaginary dimension of media productions, in an approach borrowed from gender studies[1]. It therefore seems worthwhile to propose an analytical

Chapter written by Nicolas OLIVERI.

1 See the journal *Genres en série, cinéma, télévision, médias*, Presses Universitaires de Bordeaux.

deconstruction of the mechanisms behind this idealization of post-humanist thought, in which the power of social representations plays a central role in the study of human–machine fusion through the prism of AI, both on the part of directors–creators and spectators–receivers.

1.2. From Digital Humanities to Robotic Humanities

For some years now, the term *Digital Humanities* has been particularly popular in French research, particularly in the fields of Computer Science and Social Sciences. As for the Information and Communication Sciences, they seem to be legitimately equipped conceptually to take an innovative, distinctive and interdisciplinary look at this polyform research object, the limits of which have yet to be defined, given the plethora of issues at stake. Some authors have attempted to define the field of Digital Humanities, drawing up an "inventory of French research in the international context" (Dacos and Mounier 2014, author's translation). Numerous scientific journals[2] have emerged more or less recently, proposing an editorial in line with this concept. Academic literature has also positioned itself on this promising subject, giving rise to a multitude of books, thematic issue[3], as well as colloquia, conferences and other study days dedicated to DH[4]. More recently, work in the information and communication sciences has proposed "questioning the Digital Humanities" (Paquienséguy and Pélissier 2021) through a theoretical, epistemic, ethical, political and methodological approach, in order to grasp the many issues at stake, particularly in terms of training students in the Humanities and Social Sciences. The interest in Digital Humanities seems to be shared, heuristic and perennial.

And yet, the question of what comes next, the need for a more in-depth approach and analytical framework specific to the Digital Humanities, is already being raised. This is the proposal that will be developed here, by highlighting the robotic dimension of Digital Humanities, as human–machine interactions continue to grow and strengthen, ultimately leading humanity towards a fusion that appears increasingly inescapable, and for some, largely

2 See: *Journal of Digital Humanities*, *Digital Humanities Quarterly*, *Études digitales*.

3 Bonaccorsi, J., Carayol, V., Domenget, J.-C. (2016). *Humanités numériques et sciences de l'information et de la communication*. *RFSIC*, 8 [Online]. Available at: http://rfsic.revues.org/1778.

4 Digital Humanities.

desirable[5]. This future of "the uncertain individual" (Ehrenberg 1995) should not, however, be seen as a foregone conclusion that human beings are doomed to become machines. This is not yet a reality. For the time being, it is essentially the cultural industries and certain biotechnology lobbies that are fueling the circulation of cultural and ideological flows aimed at legitimizing human–machine fusion. Nevertheless, there are many culturally identified forms of resistance, which bear witness to the hold of culture over technology, demonstrating that the latter, "however powerful, remains dependent on strong cultural variables" (Oliveri 2011, author's translation). Despite these observations of distancing, the weight of the imaginary remains decisive in the nature of the relationship that unites human and machine, partially explaining the prevailing technophilia when it comes to imagining a human being, technically augmented by technology.

1.3. Transhumanism as evidence of a diminished humanity

As an innovative and promising topic in a public and scientific debate exacerbated by sudden notoriety, the theme of "augmented human" (Kleinpeter 2013) has, for over a decade now, given rise to a number of highly contradictory accompanying discourses. A current of thought such as that of transhumanism, for example, supported by a willingly technophile media environment, relies on a particularly marked idealization of technology, in order to enunciate a desirable convergence between the organic and the technological. One of the core points of this imaginary seems to lie at the heart of various artistic currents, which actively feed it in terms of social representations. Patrice Flichy rightly points out that "the study of the technical imaginary shows that it always has two functions: to construct the identity of a social group or society, and to provide resources that can be directly reinvested in the preparation and implementation of projects" (Flichy 2001, author's translation). The resources evoked by the sociologist can then be understood as the spiritual material permitting the passage between a current representation and its future concretization. Ray Kurzweil, a computer scientist by training, theorist and media figure of transhumanism, agrees, considering "that the pace of technological change is accelerating, and that the next fifty years will see not only radical technological advances, but also a technological singularity, an inflection

5 This is the point of view defended by transhumanists in particular.

point that will change the very nature of man" (Kurzweil 2005, author's translation).

The projectivist aspect of this type of discourse draws intrinsically on a fictional universe, as it is indeed a speculation on the future, where the imaginary appears particularly attractive in order to determine a future, which by nature escapes us. Furthermore, transhumanism, beyond its fantastic promises, is above all exposed to what Patrice Flichy has identified in his work as a "mask-ideology" (Flichy 2001). In other words, the inability of a utopia to lead to a concrete, viable, tangible project, which is ultimately only of the symbolic order. In a more contrasting approach to the future of humankind, posthumanism seems to question this issue in greater detail. The temptation to use technology to compensate for the physiological limits of a human being who is by nature diminished is pressing. Behind the proselytizing of some, or on the contrary the distance observed by others, artistic culture bears witness to the propensity of futurologists, roboticists, designers and other engineers to work directly or indirectly on the development of this "technological singularity" (Kurzweil 2005). To do so, they consciously or unconsciously draw on an imaginary world of technology and robotics, the contours of which artists (of all genres) have already shaped through their many works of fiction.

In this respect, American cinema and the world of Japanese anime in particular are the art forms that have best given shape to the fantasy – finally realized – of the human–machine fusion. For example, Figure 1.1 shows the main character of the *Robocop* film series, the first part of which was directed in 1987 by Dutchman Paul Verhoeven. This futuristic film and its sequels (1990 and 1992) depict the daily life of a Detroit policeman assigned to the city's slums, accompanied by his partner. Seriously wounded during a traffic stop and left for dead by unscrupulous hoodlums, he benefits from a special scientific crime-fighting program designed to rid the city of thugs once and for all. He becomes Robocop, "50% man, 50% machine, 100% cop"[6]. This feature film is an opportunity for the director to criticize a society in which violence is omnipresent, accentuated by the complacency of the media and advertising[7]. In this future, individual freedoms are tightly controlled by a strict security policy. The role played by machines in these films is highly ambivalent, as they appear both as a technical advance

6 Tagline used on the film's poster at the time of its theatrical release.

7 A recurring theme in Paul Verhoeven's filmography.

necessary for maintaining order, and as a constant threat to turn against and eliminate their creator.

Figure 1.1. *Robocop. For a color version of this figure, see www.iste.co.uk/debos/ethics.zip*

Figure 1.2. *Rex, from Rex Bionics. For a color version of this figure, see www.iste.co.uk/debos/ethics.zip*

We noted the many similarities with Figure 1.2, which shows the lower part of an exoskeleton[8] produced by *Rex Bionics* of New Zealand. The metal-like covering, the knee joint and the specific shape of the hip, attest to a desire for this walking aid, designed for people who have accidentally lost the use of their lower limbs, to echo a robotic and fantastical imaginary world. The analogy with Verhoeven's film is all the more perceptible symbolically, since it is indeed a question of a human being whose life has been physically destroyed, but technologically saved. Numerous works of fiction, mainly Western, are based on this narrative framework, which has become generic over the decades. It is a form of glorification of technology, in which it comes to the aid of a woman or man who would be condemned to die without the massive use of technology. The "repaired" individual then puts the new functionalities and/or powers they now enjoy at the service of the community[9].

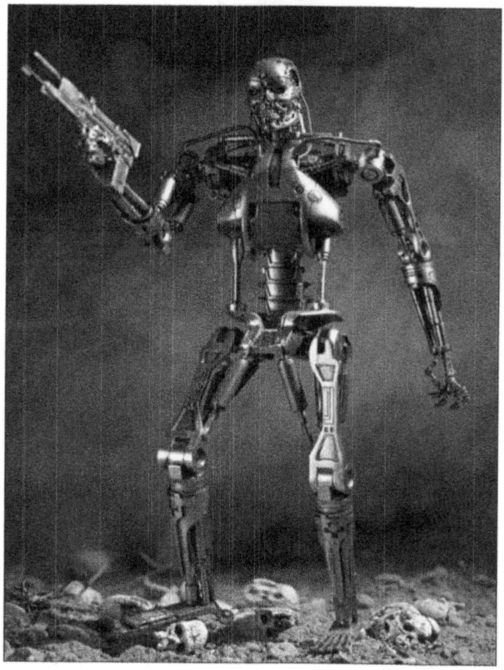

Figure 1.3. *T-800. For a color version of this figure, see www.iste.co.uk/debos/ethics.zip*

8 Known as Rex.

9 Other examples include the American series *The Six Million Dollar Man* (1973) and *Bionic Woman* (1976).

This principle of robotizing technology for human repair can also be seen in Figure 1.3. It shows the main character in the *Terminator*[10] film series, played on screen by Arnold Schwarzenegger. It is a *T-800*[11], here deprived of its flesh covering, designed to make it undetectable to humans. In fact, it is a military cyborg made of metal and circuit boards, covered with living tissue and a human-like epidermis. It is interesting to compare it with Figure 1.4, a walking assistance exoskeleton[12] developed by the American firm *Ekso Bionics*.

Figure 1.4. *Ekso Hope, by Ekso Bionics. For a color version of this figure, see www.iste.co.uk/debos/ethics.zip*

10 A science-fiction film franchise, featuring people who have joined the resistance against a super-powerful computer named Skynet, determined to turn machines against humans in order to protect themselves from the danger of being disconnected by them.

11 Fictional character from the Terminator franchise. Cyborg programmed to eliminate human beings.

12 Called “Ekso Hope”.

Figure 1.4, available on the manufacturer's website [13], is a barely concealed reference to a very specific cinematic imaginary (that of the Terminator in this case), where even the photographic framing of the product available for sale seems to mimic a deliberately jerky gait, which generally characterizes bipedal robots as they move in films or cartoons. This example is all the more interesting in that it cleverly hijacks the image of a ruthless killer-robot, in favor of an exoskeleton that effectively aims to mechanically assist physically diminished individuals. A bit like the *Terminator* in the second part of the series, after its system has been reprogrammed and its initial evil objectives erased.

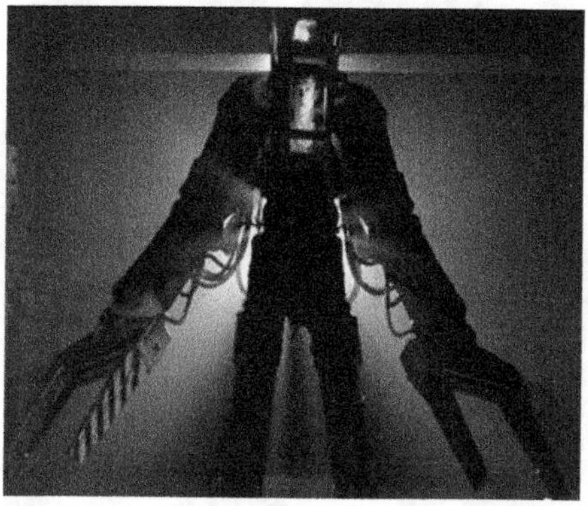

Figure 1.5. *Aliens (1986). For a color version of this figure, see www.iste.co.uk/debos/ethics.zip*

Sometimes, it is the roboticists and engineers themselves who draw directly on the vivid imagination of film-makers to design their scientific projects. Figure 1.5, for example, shows the mechanical contraption used by American actress Sigourney Weaver (reprising her role as Ellen Ripley) in the film *Aliens*[14], when she is called upon to fight the alien queen-pounder in an epic final scene. The exoskeleton project[15] (see Figure 1.6) proposed by the Japanese company Activelink is a near-faithful transposition of the motorized exoskeleton[16] imagined by film-maker James Cameron for his

13 See: www.eksobionics.com [Accessed March 26, 2021].

14 The sequel to *Alien* (1979), *Aliens* (1986) tells the story of a group of Marines who have come to explore a space base of Earth colonists, but with whom contact has suddenly been broken.

15 Called *The Dual-Arm Power Amplification Robot.*

16 Called a *Power Loader* in the film.

fantastic film, where fictional culture and technological reality meet unequivocally.

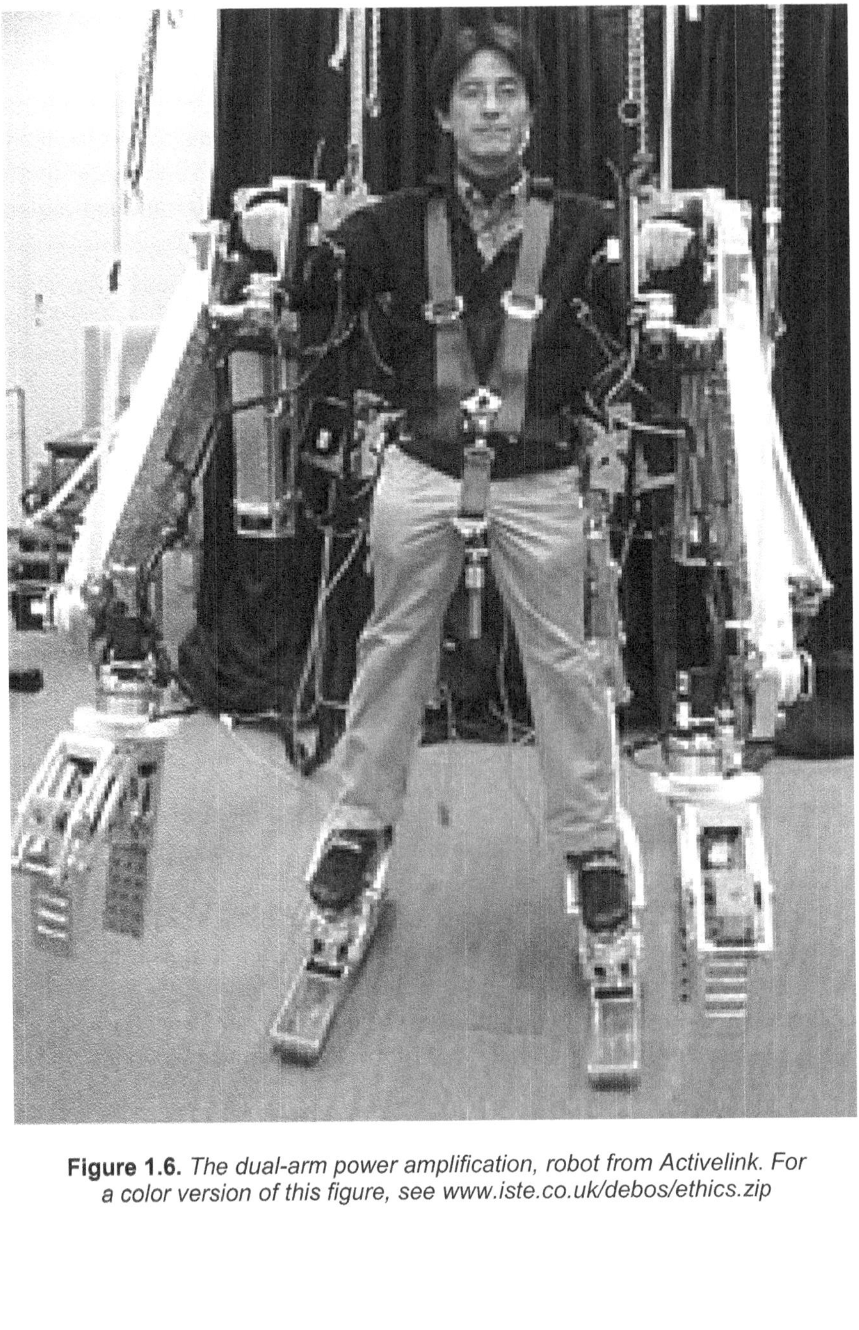

Figure 1.6. *The dual-arm power amplification, robot from Activelink. For a color version of this figure, see www.iste.co.uk/debos/ethics.zip*

Figure 1.7. *Bubblegum crisis. For a color version of this figure, see www.iste.co.uk/debos/ethics.zip*

The influence of artistic culture on recent human augmentation projects is also palpable through the prism of Japanese anime. This is particularly the case in Figure 1.7, which shows one of the main characters in a series that was very popular in the late 1980s in Japan, and in the 1990s in France, entitled *Bubblegum Crisis*[17] (AIC 1987–1991). The character is a female, dressed in battle armor. The analogy with the HAL exoskeleton project[18] (see Figure 1.8), developed by the Japanese company Cyberdyne, is very close, right down to the smallest detail. Whether it is the color scheme (white and blue) or the perfectly smooth appearance of the armor, the similarities between a production of the mind and an industrialized mechanical product are highly convincing.

Figure 1.8. *HAL from Cyberdyne. For a color version of this figure, see www.iste.co.uk/debos/ethics.zip*

Generally speaking, Japan remains a fertile observation ground for measuring the scope of robotics within industrial imaginations. Recently, many life-size reproductions of famous robots (Gundam and Evangelion),

17 This Japanese animated series follows the adventures of four young girls, clad in battle armor, as they battle against an enigmatic multinational corporation whose aim is to extend its influence over the world through the use of cyborg soldiers.

18 HAL is an acronym of hybrid assistive limb.

sometimes even animated, can be seen in certain cities[19] of the Japanese archipelago.

A further demonstration of the tenuous relationship between fiction and reality is provided by the South Korean project METHOD-1[20], developed by Korea Future Technology (see Figure 1.9). This prototype is, apart from a few details, an exact replica of the *mecha*[21] (see Figure 1.10) used by the main antagonist in the final scene of James Cameron's 2009 film *Avatar*[22]. The similarities are particularly striking in the cockpit used to house the *mecha*'s pilot.

Figure 1.9. *Avatar. For a color version of this figure, see www.iste.co.uk/debos/ethics.zip*

19 Notably Yokohama, where an 18-meter-high version of the famous Gundam robot (model RX-78-2) has been on view since December 19, 2020.

20 The *mecha* is 4 meters high and weighs 1.5 tons.

21 Refers here to a humanoid type of robotic armor, as well as the related literary genre.

22 Unlike *Aliens* (1986), the *mecha* in *Avatar* (2009) is designed for military, not civilian, use.

Figure 1.10. *METHOD-1 from Korea Future Technology. For a color version of this figure, see www.iste.co.uk/debos/ethics.zip*

Last but not least, the imaginary sometimes interferes not in the design of the finished product developed by engineers, but in its commercial name. Figure 1.11, for example, shows a hand prosthesis called *Luke Arm*[23]. This name refers directly to the character of Luke Skywalker, the main protagonist of the world-famous intergalactic saga *Star Wars*. During Episode V, *The Empire Strikes Back*, directed by Irvin Kershner in 1980, the young hero loses his hand, sliced clean off by Darth Vader's lightsaber.

23 Produced by the UK company Mobius Bionics.

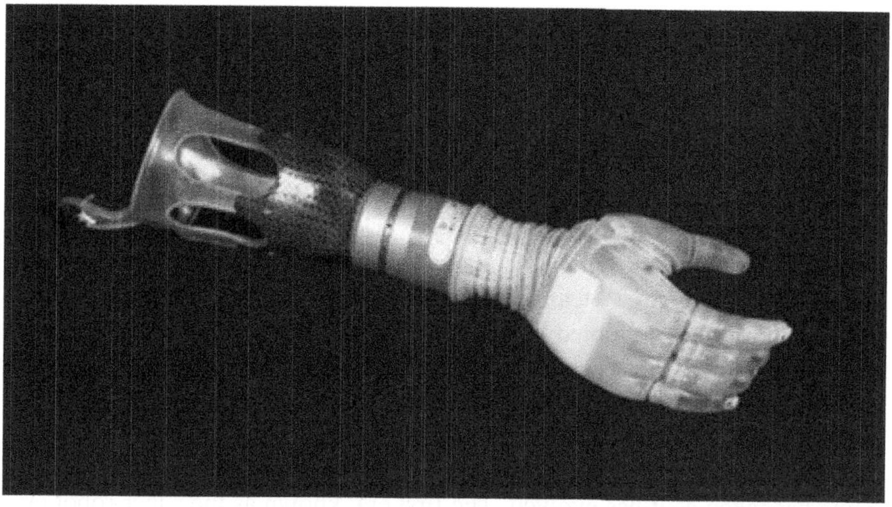

Figure 1.11. *Luke Arm. For a color version of this figure, see www.iste.co.uk/debos/ethics.zip*

At the start of Episode VI, *Return of the Jedi*, directed by Richard Marquand in 1983, technological advances enable Luke to recover the full functionality of his forearm, as shown in Figure 1.12, taken from the film. So, the six examples presented seem to support a particularly active theory of the imaginary, especially in the context of the current development of devices for human repair or augmentation.

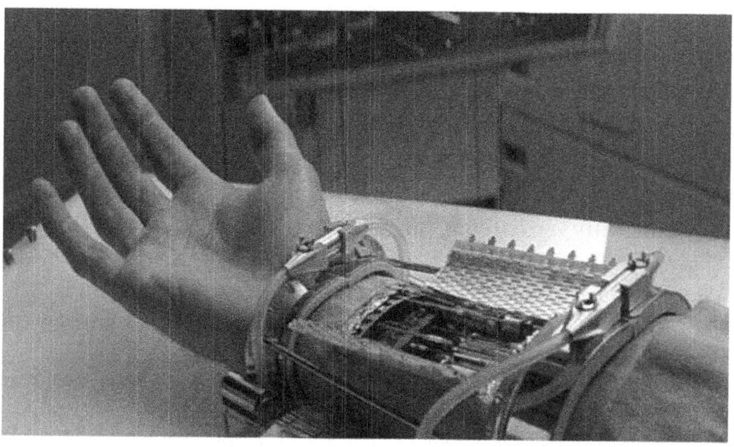

Figure 1.12. *The empire strikes back. For a color version of this figure, see www.iste.co.uk/debos/ethics.zip*

1.4. Thinking of *mecha* as an artifact of human–machine fusion

The fusion between human and machine varies from culture to culture. This is particularly evident in Japan, where the *mecha*[24] movement has historically been very popular. Previous studies [25] have shown how, culturally, Japan has developed a close and privileged relationship with the robotic environment and the imaginary world that stems from it. Trivially speaking, the *mecha* trend essentially refers to a theme specific to the science-fiction genre, featuring mechanized combat armors, often anthropomorphic and piloted by human beings. Although not widely seen in the US[26], with a few exceptions, this theme was very much in vogue in Japanese manga and *anime* from the early 1950s onwards. Even today, the genre remains attractive, although it seems to be running out of steam due to a lack of plot renewal. What remains remarkable, however, is the direction this theme has taken as it has evolved. The first era, known as the super-robot era, featured robots whose performance in combat was totally unrealistic[27]. Generally, they were piloted from the inside by a teenager with a reserved temperament, but who, by merging with the machine, revealed themself to be a true war hero. The Manicheanism of the scenarios in this sub-genre of the *mecha* current becomes a leitmotif, and the filiation between the robot and the teenager very much in evidence. In a way, the message conveyed is that the weak can become the strong, once they become one with the machine. After decades of intensive exploitation, the genre would eventually wear itself out and weary the public, in demanding need of renewal. It was against this backdrop that the real-robot[28] sub-genre emerged in the late 1970s. In concrete terms, the aim was no longer to show robots as invincible machines capable of wiping out entire armies. On the contrary, this sub-genre takes into account the robot's technical limits. The robot is treated as a simple machine, a weapon of war, with its imperfections, cumbersome operation and mechanical constraints. The plots of the series in the real-robot sub-genre, are therefore more realistic and fit into more adult, darker scenarios, generally involving armed conflict on a planetary or even

---

24 As a reminder, *mecha* is a contraction of the English term mechanics.

25 Oliveri, N. (2014). Analyse simondienne de l'otakisme et du robotisme japonais. *RIHM*, 15(1).

26 The *mecha* genre first appeared in the Marvel universe, with the Mandroids, military-piloted combat armors created in 1971.

27 Very popular in France, *Goldorak* (Nagai 1975) is the archetype of the genre.

28 Not least with the arrival of the *Mobile Suit Gundam* series (Tomino and Yatate 1979).

interplanetary scale. The emptiness of war is also very present. The ensuing success can be explained by the *mecha* genre's ability to renew itself, while aging with its core readership.

In recent years, the *mecha* trend has spread beyond Japan's borders to flood the world, mainly through blockbuster American movies featuring robotic armors designed to save humanity[29], assist humans in the face of external threats[30] or destroy them[31]. Last but not least, *mecha* has also penetrated the world of video games[32], where this theme is currently over-exploited by game publishers. The *mecha* trend thus offers an attractive and stimulating reading of a human–machine fusion, an amalgam of past and future representations of robotic technology. As for Japan, "the robot appears successively as an incarnation of pre-war slogans, a symbolic revelation of defeat by Western technology, a magnified double of the hero and a protective matrix for the outside world. Increasingly realistic, the graphics and control interface follow the evolution of robotic, digital and genetic techniques" (Suvilay 2004, author's translation). In terms of cultural production, "the robot is the medium for the fantasies and fears of a society faced with increasing technological progress, whose evolution is not accompanied by ethical debates on the place of the human being, which the machine calls into question. Science fiction thus becomes the laboratory where all consequences can be tested" (Suvilay 2004, author's translation). More than ever, the numerous testimonies to the trace of the imaginary in technology impose the idea that culture (fiction) interacts with technology (robotics). So, whether it remains a secular utopia or a finalized project, the broadened concept of posthumanism (via transhumanism and/or robotics) remains dependent on a strongly marked imaginary of technology, which ultimately steers it towards the more appropriate denomination of robotic humanities for the issues raised here. Like the development of the Internet, Patrice Flichy notes that the:

> The cyber-fantasy is not linked to a specific project, nor is it aimed at a particular audience. It is a complete fantasy construct that encompasses all aspects of this new digital society:

29 See *Pacific Rim* (del Toro 2013).
30 See the film franchise *Transformers* (Bay 2007).
31 See *Avatar* (Cameron 2009).
32 Some video games with *mechas*: *Halo, Mass Effect, Exteel, Shogo, Mobile Armor Division.*

> individual life, the relationship between mind and body, the management of society at micro and macro levels, and the production and distribution of wealth. Although all the components of this cyber-fantasy have never been brought together in a unified discourse, if we were to engage in this work of synthesis we could construct a narrative analogous to the great utopias of the seventeenth and eighteenth centuries, describing the workings of another world (Flichy 2001, author's translation).

Robotic humanities seem to be built on similar ideological foundations. And while certain technological advances[33] do indeed seem to confirm the determination of many of the actors involved to tip humanity towards a hypertechnological destiny, culture (still) seems to dominate technology. The latter, despite its galloping hegemony, conforms to an imaginary that remains human, which ultimately subjugates its role within today's societies, still leaving humankind with the prerogatives of their technological future. Finally, it is also worth pointing out that the myth of the robot, the rival brother of the human being, attracts as much interest and support from certain researchers as it does fear and apprehension[34] from a large part of the general public, who are reluctant to allow themselves to be robotized without reacting, and yet are always quick to recognize technology as unquestionable progress.

1.5. Representations of artificial intelligence in Japanese anime

Themes such as robotics, dystopia, computer networks, cyborgs and advanced technologies have abounded in Japanese anime since the 1980s, if not earlier. It would be a tall order to list them all here. However, the theme of artificial intelligence in Japanese fiction, a form of updating these past questions, has recently aroused particular interest at the heart of certain works

33 Watch Philippe Borrel's documentary, *Un monde sans humains?* ARTE France/Cinétévé, 2012, 52 min.

34 See the Uncanny Valley Theory (Mori 1970). Formalized by Japanese roboticist Masahiro Mori in 1970, it states that the more similar a humanoid robot is to a human being, the more its imperfections seem unbearable to look at. As a result, a robot's artificial appearance seems to be better accepted by individuals. If the degree of realism is high enough, the theory then predicts that a person will have managed to get past the disturbing metaphorical valley, and finally accept the robot in front of them.

of anticipation or science fiction. Here, we take a look at one such work, *Psycho-Pass*. *Psycho-Pass* was created in 2012 by Katsuyuki Motohiro and Naoyoshi Shiotani, and produced by Productions I.G. Studios, renowned for their animated adaptation of the world-famous manga *Attack of the Titans* (*Shingeki no Kyojin* in the original version). *Psycho-Pass* is first and foremost a rather successful blend of genres. A rather standardized technological crime thriller at first glance, its plot is based on the vision of a Japanese society which, while ideal in theory, is one in which individual freedoms are closely monitored and then controlled. The theme of dystopia (inverted utopia) is very much present, and quickly moves towards exposing the dysfunctions of such a social model, which is supposed to ensure the security of all. Indeed, given the political failures of the various governments of the past, the city of Tokyo (the fictional universe is set in 2112) is now managed by an artificial intelligence called Sybil, capable of determining the crime rate of each inhabitant of the megalopolis. A special police unit (Security Bureau) is tasked with intervening whenever a citizen's crime rate reaches a certain threshold, predefined by Sybil. By threatening a suspect with a pistol called *Dominator* (a polymorphic weapon), the crime coefficient is then quantified and determines the procedure to be followed by the police officer: neutralization of the target, arrest and detention in a re-education center or, more rarely, physical elimination of the suspect if their coefficient proves too high in relation to the criteria set by Sybil. This "intelligent" system is designed to guarantee social peace, by preventing or eliminating all forms of threat to fellow citizens. As the plot progresses, the liberticidal nature of expedient justice and a binary vision of society ("good submissives" vs. "bad rebels") are challenged by various characters, some of whom are asymptomatic criminals and therefore undetectable by Sybil, as well as by the police themselves, who end up questioning this type of "perfect" society and their role within it.

The originality of this series also lies in the quality of its dialogues, in which certain characters refer to numerous classic literary works (mainly European), rather unusual in Japanese animated series. Over the course of the various seasons, extracts from the following works are quoted or mentioned in the course of the narrative: *1984* (George Orwell), *Heart of Darkness* (Joseph Conrad), *The Republic* (Plato), *Beyond Good and Evil* (Friedrich Nietzsche), *Discipline and Punish* (Michel Foucault), *Principles of Philosophy* (René Descartes), *Neuromancer* (William Gibson), *Faust* (J. W Goethe), *Introduction to the Principles of Morals and Legislation* (Jeremy Bentham), *Pensées* (Blaise Pascal), *The Wretched of the Earth* and *Black Skin, White Masks* (Frantz Fanon), *Economy and Society*

(Max Weber), *A Discourse Upon the Origin and the Foundation of the Inequality Among Mankind* (J.J. Rousseau). What sets this series apart is its "cultivated" approach to the question of a future in which artificial intelligence would occupy a central place in the destiny of a given society. The question of identity remains central, for both humans and machines. What remains most remarkable is the series' authors' denunciation of a world where AI has taken political control of a state, in this case the Japan of tomorrow. However, there is no real Manichaeism here, since the series' protagonists (humans, i.e.) are, to varying degrees, responsible for the world they have shaped through the massive use of technology.

Generally technophile in its uncomplicated approach to technology at the heart of culture, Japanese anime in principle values the rapprochement of the human and the artificial, through an expected scenario in which an introverted young boy in search of his father, will reveal his true power by associating with a robotic figure of paternal substitution[35]. This type of storytelling is well established in Japan, and meets the script specifications of numerous works that have gone down in history. A priori, AI does not seem to arouse the same enthusiasm among *mangakas*[36], who highlight its shortcomings and the dangers it represents for democracy and individual freedoms. A cross-cultural approach, analyzing the Anglo-Saxon, European and Japanese visions, seems to be the consensus for identifying AI as a potential rival to humans, capable of supplanting them in the more or less near future. The robot, on the other hand, is far more ambivalent in its role as competitor, its anthropomorphic stature paradoxically lending it human characteristics (head, trunk, arms, legs). It remains a physical, tangible embodiment which human can still intervene in, destroying it if need be. The AI, on the other hand, is everywhere and nowhere, wired, "reticular" (Musso 2003), totally autonomous, decentralized via networks, and ultimately omnipotent. An invisible enemy, in short, against which humankind is unable to fight effectively. What is more, the current over-mediatization of AI and the almost daily presentation of its performative dimensions[37] seem to fuel fears of an uncertain future for the prowess of these technologies, over which human could ultimately lose control. Japanese fiction, such as

35 This narrative pattern, known as *nekketsu* or "hot-blooded", is at the heart of many Japanese animation plots for boys and young men (*shonen manga*).

36 Authors of *manga*.

37 Examples include conversational agents such as ChatGPT, Bard, Bing AI and Apple GPT, which are as powerful as they are worrying, and AI programs such as Midjourney, which can generate images from text data.

traditional Franco-Belgian comics[38], shares this alarmist vision of a political, ideological, economic or social recuperation orchestrated by AIs, as part of a Luddite vision[39] of technical innovation, where technophobia is never far away. At times an assertive antagonist, at other times a narrative pretext for an author's partisan affirmation, the scripted treatment of AI in fiction remains generally associated with the denunciation of its shortcomings, the societal dysfunctions it provokes, its misappropriation for criminal purposes, etc. Rare are the works in which AI is used as a pretext for an author's partisan affirmation. AI rarely plays a positive role in the society in which it is deployed.

1.6. Conclusion

The question of artificial intelligence remains particularly prevalent in fiction, whether literary or animated, Western or Eastern. A current reconfiguration of past questions about the future of technology and its gradual, inescapable integration into human daily life, this theme continues to provide abundant fodder for a current of thought charged with mistrust of the machine, as soon as it imposes itself on humans as a "machinic double" and likely candidate for replacement. This feeling of attraction–repulsion for AI is both expected and unthinkable. Indeed, "by simulating intellectual functions that seem properly human, the computer has become, in the eyes of certain experts of artificial intelligence […] a reflection of the mind, and then a model for thinking about intelligence. […] The computer is promoted to the rank of brain, and the brain assimilated to a thinking machine through a rhetorical game that is effective in contemporary social imaginaries" (Le Breton 1999, author's translation).

Recent examples clearly identify the principles of circularity at work between creativity and artificial intelligence. Indeed, it is now possible to

38 Titles such as *Carbon & Silicium*, *IAN*, *Sentient*, *Magnus*, *Flesh Empire*, *Intelligences Artificielles*, *Miroirs de nos vies*, offer a scenario in which AI is presented as oppressive, protective, uncontrollable, liberticidal, in search of identity or freedom.

39 In the 19th century, during the Industrial Revolution, England experienced violent social conflicts between wool and cotton shearers, weavers and knitters, and employers and factory owners. The mechanization of manufacturing processes, and in particular the arrival of looms, provoked a series of revolts and machine sabotage in 1811–1812.

entrust an AI with the task of continuing the series of a deceased author. This was the case with Osamu Tezuka[40], who, several years after his death in 1989, had an AI continue his work[41] based on the universes, themes, characters, plots and dialogues from his previously published works. Or when AI, the subject of reflection, becomes AI, the object of creation. These deceptively new issues were already being echoed in particularly pessimistic tones in the mid-1970s. "Do machines think? Will they one day be more intelligent than man, their creator? Will our biological civilization be succeeded by a civilization of supercomputers" (Ribes and Biraud 1975, author's translation). Clearly, the barrier between humanity and "machineness" is becoming increasingly blurred, reinforcing the fear that human is becoming "machineized", while at the same time giving the machine the opportunity to become increasingly humanized by contact with humans. The loss of control is not just political, economic or ideological, as the *Psycho-Pass* series reveals. What is even more worrying is the ontological loss of control, where the very erasure of the human being is at the heart of the many plots that deal with this subject. Against this ominous backdrop, fiction writers are both eager to discuss the issues at stake, but already seem resigned to the likely dark future of their works.

1.7. References

Dacos, M. and Mounier, P. (2014). *Humanités numériques. État des lieux et positionnement de la recherche française dans le contexte international.* Institut français, Ministère des Affaires étrangères pour l'action culturelle extérieure de la France, Paris.

Ehrenberg, A. (1995). *L'individu incertain*. Hachette, Paris.

Flichy, P. (2001). *L'imaginaire d'Internet*. La Découverte, Paris.

Kleinpeter, E. (ed.) (2013). *L'humain augmenté*. CNRS Éditions, Paris.

Kurzweil, R. (2005). *The Singularity is Near: When Humans Transcend Biology*. New Penguin, New York.

Le Breton, D. (1999). *L'Adieu au corps*. Métailié, Paris.

40 Artiste prolifique (mangaka, animateur, *character designer*, producteur et scénariste), Osamu Tezuka est l'un des grands noms du manga à travers le monde, principalement connu pour ses séries *Astro Boy*, *Le Roi Léo* ou encore *Black Jack*.

41 Le manga *Paidon*, publié en 2020 au Japon, a été créé par une IA à partir de 65 œuvres du mangaka disparu.

Musso, P. (2003). *Critique des réseaux*. PUF, Paris.

Oliveri, N. (2011). La cyberdépendance : un objet pour les sciences de l'information et de la communication. *Hermès, La Revue*, 59, 167–171.

Paquienséguy, F. and Pélissier, N. (2021). Questionner les humanités numériques : positions et propositions des SIC [Online]. Available at: https://www.sfsic.org/wp-inside/uploads/2021/06/questionner-humanites-numeriques.pdf.

Ribes, J.-C. and Biraud, F. (1975). *Le dossier de l'intelligence artificielle. Demain les machines penseront*. Fayard, Paris.

Sulivay, B. (2004). Robot géant : de l'instrumentalisation à la fusion. *Belphégor*, 3(2), hal-04530271.

2

The Boundaries of Technology and the Future of AI

2.1. Introduction

In the contemporary era, we increasingly interact with non-human entities within complex digital environments. The fusion of AI-generated content and human activity blurs the lines of origin, and challenges our perceptions of authenticity, communication, authorship and more. Furthermore, AI's precision and processing capacity go beyond human language, pushing and the traditional boundaries of technology and communication. This transformation compels us to reimagine the possibilities of AI and their ethical implications, which is the focus of this chapter. This phenomenon is a departure from traditional human-centric communication, as Braidotti eloquently explains in his exploration of posthuman theory (Braidotti 2013). Braidotti's work encourages us to reassess the limits of technology, as we embrace interactions with artificial intelligence, autonomous systems and even a more-than-human world. These ideas encourage us to expand our repertoire of communication, envisioning a future where exchanges between, and even through, humans and non-human actors become commonplace. As we navigate this evolving landscape, innovative and ethically grounded approaches to these new forms of dialogue take center stage, promising to redefine our relational frameworks in as yet unexplored ways.

While both disruptive and new, the widespread adoption of AI-based generative technologies is deeply entangled in a long-standing debate. It is not possible to limit this discussion to a simplistic framework. It has

Chapter written by Juan Antonio CARRASCO.

therefore become imperative to move away from technopastoral and technophobic narratives (Sadin 2016, 2018). We argue that a fuller exploration of the frontiers of technology allows us to uncover the ethical opportunities and challenges of AI generative technologies and their diverse impacts. Previous chapters in this book have accurately traced the light and shadows of our engagement with AI-based technologies. This pendulum of stories swings between passionate enthusiasm and deep skepticism, encapsulating the plexus of questions between technological change and its potentially dehumanizing consequences. This dynamic extends beyond academic debate, permeating cultural expressions through cinema, literature and animation, mobilizing the imagination around the cutting edge of technological innovation. These representations, reinforced by recent advances in AI, are the vectors of our deepest hopes and fears.

As this chapter proposes, rather than adopting a supposedly equidistant stance between these narratives, a step back allows us to re-examine the phenomenon: information communication departs from traditional human-centered models by prioritizing functional and practical outcomes. However, what are the ethical and philosophical issues involved? In essence, dialogue between humans and non-humans – whether artificial intelligence, animals or ecosystems – refers to non-human forms of communication.

To explore this idea further, it is essential to examine concerns related to the normalization of communication and cognitive and perceptive capacities. This examination could involve assessing how generative technologies, while intended to support democratic goals, could potentially be used to distort and manipulate information. Critically assessing these implications is essential to advancing democratic debates. This chapter proposes a reflection on three key ideas for navigating the limits of technology and the ramifications of artificial intelligence. The central structure of this reflection is how Derrida's politics of the archive intersects with Marcuse's and Heidegger's exploration of the essence of modern technology, creating a dialogue in which an understanding of technology in the contemporary world is looked at. Such an examination allows us to better understand the ethical implications of the rapid evolution of AI-generative technologies in the field of communication. This perspective invites us to rethink our approaches to interactions, collaborations and coexistence with the myriads of non-human entities that share our environment beyond mere technologist and millenarian discourses, considering its broader implications for democracy and societal structures.

2.2. Generative AI and exponential technology

Exponential technologies, characterized by their rapid and disruptive growth, are redefining the boundaries of innovation and societal impact. Generative artificial intelligence falls into this category because of its ability to evolve and perfect itself at a pace that far exceeds traditional technological development rates. This form of AI, with its sophisticated algorithms that learn and create new content from vast data sets, is showing exponential growth in its applications and capabilities. However, the ubiquity of digital technologies and their incessant transformations can often prevent us from understanding the historical context and conditions of their emergence. Our decisions continually shape the information systems on which we depend. At the heart of digitization lies the effort to integrate a wide range of logical problems and operational inquiries across diverse domains, providing a medium of "materiality" and encompassing a spectrum of ethical issues.

Moreover, digitization represents the confluence of logic and technology in a broader assessment, endowing the various challenges with a tangible logical framework. As Kittler notes in Gramophone, Film, Typewriter (1986), the analytic principle of computation consists of breaking down any problem, question or process into increasingly simple elementary operations. This principle of analytical decomposition paves the way for the fundamental progression of binary code in computing. Thanks to the binary language of zeros and ones, the computer process has not only facilitated the resolution of complex problems by breaking them down into simpler elements, but has also established a logic through which calculations acquire shared metrics. This logic serves as a framework for thinking, enabling calculations that promote discussion and, consequently, a form of shared reference and measurable reality. The primary division between computing units and storage memory, coupled with a register, establishes the foundations of standard computer architecture (Boullier 2016). Moreover, cryptography is at the heart of the founding moments in the development of computing and artificial intelligence. The vertiginous transformations observed over the last few decades are also attributable to advances in miniaturization technology. These developments echo the ideas expressed by Bijker in *The Social Construction of Technological Systems* (1987), where he illustrates the interaction between technology and society, emphasizing the role of social choices in technological evolution. The rapid technological changes brought about by digitization plunge us into a state of uncertainty, where traditional social practices are extracted from their local context and restructured in an indefinite time and space, underlining the profound impact of digital technology on societal structures and individual existence (Giddens 1990).

Generative AI distinguishes itself by providing completely new results without human supervision, thanks to sophisticated linguistic models adapted to interaction and problem-solving. This linguistic model manipulates symbolic elements and obtains information through recursive and intensive information processing operations (Smith and Jones 2021). AI models demonstrate a structured precision in the processing of linguistic elements, which contrasts with the inherent limitations of human communication. Unlike AI's algorithmic clarity, human language thrives on nuance and adaptability and carries a degree of unpredictability and error that AI models generally do not allow. Therefore, AI systems are designed to facilitate linguistic interactions in a digital environment, generating coherent and context-sensitive discourse, even in domains requiring expertise (Doe 2020). This capacity for dialogue and discourse generation is not simply a technical achievement; it encompasses implications for diverse applications, such as autonomous vehicles, legal analysis or medical consultation, by reproducing complex patterns of human interaction (Brown 2019; Lee and Kim 2020). The dialogical how we interact with technology (Johnson dialogical dimension of generative AI highlights its potential to revolutionize how we interact with technology (Johnson 2022). However, what are the implications of generative AI technologies for our culture and communication? By bridging the communication gap between human and machine, generative AI goes beyond functional utility to influence how knowledge is constructed, shared and evolves within society (Martinez 2021). Furthermore, the implications of generative AI's ability to understand and produce human-like texts have an ethical dimension, raising important questions about the boundaries between human creativity and machine production.

As these systems become more adept at imitating human writing (Lévi-Strauss 1955; Derrida 1967), this is relevant to democracy and political issues. Furthermore, the deployment of generative AI in sensitive contexts, such as news creation or research, introduces the risk of misinformation or undetected dissemination of biased viewpoints, underscoring the need for transparent disclosure of AI's role in content creation. Ethical guidelines and regulatory frameworks must evolve to meet these challenges, ensuring that generative AI can develop human creativity and knowledge dissemination without undermining the integrity or reliability of the information ecosystem. Equally important are the ethical and philosophical considerations surrounding the impact of this technology on human interaction, as well as creativity and autonomy, sparking ongoing debate about the future role of AI in society (Nguyen 2020). As we delve deeper into the capabilities and potential applications of generative AI, it becomes increasingly clear that this technology is not simply an extension of existing digital tools, but marks a critical shift in the interface between humans and machines, posing unparalleled ethical and linguistic dilemmas (Patel and Singh 2019).

The exploration of generative AI technologies reveals that it has become difficult to distinguish humans, from artificial creation, raising critical questions about the essence of communication and language. It seems pertinent to examine the approaches of Heidegger and Marcuse. Although their contexts and concerns differ, both thinkers offer perspectives on technology that shed light on our current situation. The following analysis helps us better understand our relationship's global configuration to the technical world.

2.3. Technology as the culmination of metaphysics

Consequently, the possibility of addressing the place of technologies based on generative AI, as well as the question of ethical responsibility in the face of its various manifestations, is positioned in its connection with a historical framework concerning the state of technological progress, and its significance in the world (Heidegger 1958). This position is ambivalent, and cannot be reduced to a set of unambiguous, straightforward positions. From the 1930s onwards, the question of the essence of modern technology and how it reconfigures our relationship with technology emerged as a central theme in Heidegger's thinking, explored through a variety of works (Heidegger 1958, 1976, 1983). In the process, Heidegger explores a series of variations centered on thinkers, and the task of destroying the metaphysical conceptual tradition based on its key figures: subject, idea, form and phenomenon. This terminological elaboration acts as a breadcrumb trail through the labyrinth of metaphysics.

Heidegger conceives of metaphysics as the history of the forgetting of being through its various manifestations. In this context, we have neglected the question of difference, assimilating being to a series of ontic schemas, thus privileging a particular understanding of reality. Technology bursts onto this *continuum*, revealing itself as the culmination of metaphysics and its final realization. Throughout its trajectory, metaphysics has traversed the ring of its essential possibilities (Heidegger 1978). Therefore, in the contemporary context, reality is objectified by modern subjectivity, which transforms the world into an entirely manipulable and technicized environment.

However, Heidegger strives not to present the power of technology as an isolated or intrinsically harmful phenomenon, thus offering a more balanced perspective on human interaction and digitization. Nevertheless, in the essence of modern technology, the circle of the oblivion of being is closed. This is why, in 1953, Heidegger introduced the term *Gestell* to designate the central aspect of the technical world. Because of its typically Heideggerian composition, this concept has been retained in translation to designate the locational power deployed by modern technology with regard to nature and human realities (Heidegger 1958). The greatest danger lies in modern technology's drive towards the reduction of reality, which removes all alternatives and threatens to eliminate all memory of other ways of experiencing and inhabiting the world. This connection with Marcuse is underlined in his 1968 work. The threat comes not from the machines or technical devices themselves, but from the global configuration of our relationship to technology, which can prove a major risk. For Heidegger, the implementation of the *Gestell* entails the risk of distancing ourselves from the deepest truth, of being deaf to the original call of truth.

Nevertheless, somehow, this primary truth has revealed itself to us as the freedom to look at things, to let them manifest themselves. The openness and clarity that emerge when we find a way through the rage of technology represent the essence of truth, and the hope of interacting with nature in a way that allows us to ask new questions (Heidegger 1983). According to Heidegger, in his *Letter on Humanism*, this approach encourages us to care for nature, in contrast to "pure" technical domination. It is not just a question of waiting for a new way of thinking to emerge, but of recognizing that another way of thinking or being in the world can project new historical possibilities. Heidegger pointed out in his 1953 lecture, "The Question of Technique", that there is a notion of epochal emptiness and the possibility of a more open engagement with technology. This ambivalent character of the essence of modern technology as a destined dispatch is expressed in Hölderlin's lines: "But where the danger is, grows the saving power also."

In contrast to *Gestell*, Heidegger presents the opposite notion of producing in the sense that it can become reminiscent of thought, serenity, becoming. Suddenly, these two attitudes to nature – provocation and letting go – were precisely defined in a lecture on thought. Looking back on this fatal sequence of *Gestell* and preparing for a movement in history, but a movement we cannot at all anticipate or prepare for, memorial thinking creates a space for contemplation and analysis of the phenomenon. Therefore, the significant change in thinking that Heidegger (1958) identifies has already begun to manifest itself: we accept the presence of technical objects in our everyday environment, allowing them to occupy space not as absolute entities but as manifestations or figures of being. This position invites us to rediscover an ancient word that sums up our attitude to the vast technical world: *Gelassenheit*. This serenity in the face of things, rather than a simple passive attitude, represents a profound change in our way of thinking, even if it does not necessarily translate into a direct program for transforming history.

Against critics pointing out the ineffectiveness of this approach, Heidegger would argue that the demand for reliability is already a notion belonging to the realm of technical thought, limited by the *Gestell*. According to Heidegger (1958, 1976), no credible solution within the present framework can effectively resolve the dilemma of technique. This perspective does not advocate fatalism, but rather suggests an unintentional model of relation to the technical world, unmediated by the subjective frameworks generally used to interpret our interactions and calculations. In essence, Heidegger calls for an engagement with the world, in which no single human action can substantially alter the current order dominated by global powers; this established order irremediably influences our actions, and we are therefore subject to it. This reflection encourages us to reconsider our relationship with technology and our role in the universe in a more thoughtful, less calculating way.

Under the sign of modern technology, we do not let nature come into being. We provoke it and approach it in a way assimilated by calculation, so that its power is configured as an information system, transforming nature into a resource, a kind of static background at the disposal of the ends pursued by standardized production (Heidegger 1958, p. 31). This transformation, in which nature is reduced to a resource, is an essential feature of contemporary technology. Therefore, technology demands a surplus of techniques, suggesting that we cannot master its consequences without more techniques.

2.4. The commodification of imagination in the technical world

Herbert Marcuse, intellectual heir to Adorno and Horkheimer, as well as a student of Heidegger, is particularly interested in the problem of domination in the technical world. According to Marcuse (1968), the risk inherent in technologically advanced societies lies not so much in the possibility of authoritarian regression, but rather in their connection with a new form of social organization whose structure is intrinsically repressive. This society is totalitarian for Marcuse, but not in the sense of the fascist regimes denounced by the *Dialectic of Enlightenment* (Adorno and Horkheimer 1972). In this context, political and social coercion do not arise directly, but are mediated by subtle integration mechanisms such as consumption, leisure time, public opinion and technological development, which ensure a relative level of well-being for a majority of the population, thus neutralizing any questioning of the structural contradictions that underpin the system.

From this perspective, the trends that characterize advanced societies include the expansion of consumption in an increasingly unequal society, where the boundary between the public and private spheres is blurred, and where the individual is constrained by the parameters imposed by public opinion. Against this backdrop, the dynamics of social integration are intensifying, particularly among urban wage earners and, selectively, in areas of potential conflict. The expansion of social protection systems, which broaden political rationality and regulate exchange relations, makes it possible to improve social and working conditions by technically resolving routine tasks.

His point of view suggests that technical mastery makes it possible to adapt communication to social requirements from the most utilitarian point of view. It is no longer imagination that speaks, but reason in a reality that justifies its demands for reproduction at all costs, even in an order where the realization of the conditions of production is not possible. Imagination abdicates precisely in the face of this reality but somehow transcends and surpasses it. For Marcuse (1968), technological progress is inextricably associated with a fracture where rationalization and the progressive realization of the imagination are articulated. This dynamic leads to a world in which the traditional archetypes of horror and joy, war and peace, are completely transformed. They are no longer irrational forces of everyday life, but modern avatars, symbols of the technological domination to which we are subjected. Society shrinks as images are transformed into potentialities and historical projects, although this transformation often fails, becoming erroneous and futile outside the realm of material production and needs. This is where the imagination expresses itself, bound only to fantasy, thus losing its potential for emancipation. The fusion of technical progress with imagination, reverses the scenario of its logic and truth, diminishing the once accessible faculty of the mind. Nevertheless, this same mastery of technology brings imagination and reason closer together (today, probably more than ever), binding these historically antagonistic faculties into a fundamental dependency. With the progress of industrial civilization, the games of the imagination no longer revolve around technical eventualities or the evaluation of their realization.

Other theoretical developments illustrate a move towards the recognition of the rational and scientific nature of the imagination within the framework of mathematics, research hypotheses and the physical sciences. Psychoanalysis, as we have seen, recognizes the particular rationality of the irrational, positioning the imagination as a therapeutic force with emancipatory potential. Marcuse (1968) critically observes the process of reification of the imagination in technologically advanced societies, suggesting that we are possessed by and suffer from our creations: a new kind of digital fetishism has emerged. Understanding the depth of this dilemma, critical theory warns us that fully liberating the imagination is also a form of regression. Those with a mutilated imagination tend towards organization and destruction, highlighting a liberation that not only cultivates criticism, but also accelerates its most repressive tendencies.

Imagination thus becomes a beacon for production and a guide to a serene, distress-free existence. This liberated imagination contrasts sharply with the one shaped by domination. Freeing the imagination and fully embracing its capacity for expression implies suppressing much of what is present and unbridled in a repressive society. This regression is not simply a psychological or moral issue, but a profoundly political one, emphasizing that politics is the arena where societal institutions evolve, and where they are defined, preserved and modified. Marcuse reiterates the argument: how can individuals, limited in their freedoms and satisfactions, emancipate themselves from their controls? In this context, breaking this vicious circle is an arduous task, not because it requires creating new social institutions – societies are already in transition – refining their fundamental structures through precise planning or markets. However, critical theory cannot be propositional in indicating a program, a path of transformation; it is true that the dialectical concept that comprehends given facts, also transcends them

(Adorno and Horkheimer 1972). In this transcendence, it precisely receives its emancipatory potential; it determines eventualities and historical needs, but only practice can realize theory, and, at present, no historical practice realizes it. This realization brings us to a critical point: Does it mean that the critical imagination must abdicate? Dialectical concepts, as discussed by Habermas (1981), are once again found in their truth; they allow us to understand their situation in social reality (not necessarily to transform it). The answer lies there: as with Heidegger, any social transformation or reform program is rooted in the assumptions it seeks to overturn. Theory can only remain negative in its critique of established society; the vestiges of emancipation lie in the ability to escape the conveyor belt of non-generalizable particular interests, and to denounce their alleged rationality.

Until such time as the fundamental social forces have proven themselves, the critique of society must be rational. However, it cannot translate its rationality into terms of historical practice. What are we to conclude from all this? The liberation of inherent potential no longer allows us to adequately experiment with possible historical evolution. The eventualities that advanced industrial society has rushed to realize, by constantly developing the productive forces on a universal scale, underline the limits of critical theory in the face of the realities of this society. So, if critical theory cannot remain negative, for Heidegger, the possibility of an open relationship to the essence of modern technology lies in a way of thinking that lies somewhere between memory and a critical responsibility towards language. This alternative path offers a way forward, suggesting a nuanced approach to engaging with the essence of technology and its impact on society.

2.5. Archive policy and future democracy

In 2002, in *Echographies in Television*, Derrida explores the role of media technologies in the formation and production of events. He analyzes how reality is technically and artificially produced by information and its effects on democratic politics. This approach emphasizes that meanings are always subject to interpretation and depend on the complex interplay among the beliefs, desires and intentions of speaker and interpreter. Our interest in generative AI could illustrate how linguistic constructs can be created and manipulated, demonstrating the contingency of all forms of linguistic expression.

The deconstructivist approach shows that reality and technically produced events reveal that pure facts do not exist, as Nietzsche asserts; on the contrary, events are recorded and modified by their reception in the media. This perspective requires an analysis of how events are captured and repeated by technology, focusing on their dissolution into fiction. While producing the reality it presents to the observer and described as "artefactual" by Goldschmidt (2003), technology does not generate the event itself and cannot serve as technically established proof. This underlines the crucial role of the observer, as technical terminology can only describe the terms of an event, not the trace of the event. Thus, the critical analysis of the dismantling of the artifact provides a faithful testimony to the event.

The deconstructive analysis differs from Baudrillard's interpretation of mediated reality, which argues that events such as the Gulf War did not happen because their existence for us was purely mediated and virtual. According to Baudrillard (1995), the simulacrum of images and the manipulation of television news and reports have cancelled out the real place of the event, suggesting that we live in a continuous simulacrum rather than reality itself. Derrida notes a phenomenon similar to Baudrillard's concept of simulacra, and exacerbated, as interoperability alters the structure and character of events. Deconstruction reveals the transformation of the event into a simulacrum, which begins when the event is recognized as such, repeated by information and differentiated by itself. This process of exposure and repetition shapes the event, but each repetition alters it, preventing it from remaining identical. Derrida describes this phenomenon as "iterability", where the archived, reported and historicized event emerges transformed (Derrida 1967).

Deconstructive analysis thus advocates a kind of politics of the archive, suggesting that the event escapes its initial trajectory and fades away, as it manifests itself only in the repetition and alternation mediated by technology. An event is not recognized as such until it is interpreted in the repetitive context that gives it meaning. For deconstruction, this structural blurring or erasure of the event implies an ethical call to responsibility and memory. Precisely because the event is diluted in repetitive media images, deconstruction insists on the singularity and painstaking work of memory, underlining the importance of not letting this constitutive erasure be forgotten.

The approach to the singularity of the event, seen as "having taken place", serves as a tactic against forgetting, highlighting the present but erased elements of the process and logic of simulacra. This reflection on singularity highlights the contradiction between the transformation of the event into a simulacrum and the singularity of its occurrence as a critique. Through this contradiction, deconstructive responsibility follows the traces of the event's memory. This responsibility consists of analyzing the event's singularity and its erasure in simulacra, which is only possible through rigorous archiving. As Derrida (1995) points out, the ability to access, analyze, select, interpret and understand the manipulation that has influenced the production and circulation of events is not just a citizen's right, but a fundamental pillar of democracy. The deconstruction of the archive demands responsibility in consideration of singularity, establishing this as a right for citizens rather than merely an abstract philosophical or theoretical endeavor. Effective deconstruction occurs when the archive of events is critically analyzed, facilitating an understanding of its production and archiving methods, thereby supporting democratic processes. It is imperative that policy timely guarantees the formal rights and technical conditions necessary for access to archives, as this has a direct impact on the preservation of democratic values.

If the application of the law or delays in a specific area impede a person's right of access to archives, this constitutes a violation of that right. The evolution of archiving techniques, stimulated by modern communication technologies, can accelerate democratization processes more effectively than legal proclamations or political declarations of equality. This optimistic vision of technology must be linked to a practical deconstruction of established powers and enhance democratic outcomes, potentially accelerating democratization efforts, and challenging authoritarian tendencies.

Beyond any moral judgment on technology, Derrida reflects on how technical reproducibility transforms the essence of politics. His analyses are realistic in the sense that they neither condemn technology nor seek to block the historical transformations it irrevocably entails. Rather, deconstruction distinguishes various technical trends and processes, which sometimes confront or counteract each other, given the potential transformations on the horizon of a future democracy. We cannot juxtapose technology with an earlier state of craftsmanship, for even what came before has been shaped by technology. Technical reproducibility also creates a latent and imposing suspense. In essence, it appropriates the event, modifying the notion of living presence across the register, since every living presence is preceded and deconstructed by writing, sound, photography and film. In this spectral dance, Derrida asks whether we can understand the evolution of modern technology as a force that decomposes or exploits power. These facts return altered and different, representing events that repeat themselves, but transform beyond their original repetitions, altering the contexts they traverse.

2.6. Conclusion

This chapter discusses some of the philosophical and ethical concerns associated with the technology, and their implications for understanding generative AI. In this regard, a major concern remains the presence of cognitive biases in algorithms or the direct manipulation of information for political ends, transforming the traditional roles of "communicator" and "recipient". This transformation recalls the challenges raised initially, blurring the original lines and questioning our perceptions of authenticity and interaction. This change requires a deeper understanding of communication, encompassing the dynamic interplay between human and machine intelligence.

On the one hand, Heidegger calls for measured reflection on the essence of technology, suggesting that we cannot change our technological destiny by sheer willpower, or return to an earlier state of simplicity. However, he also warns against the unthinking pursuit of technological progress for its own sake. Similarly, though from a distinct perspective, Marcuse emphasizes the need for a robust critical imagination. These perspectives encourage a critical re-examination of the view that technology serves as an instrument for achieving predefined goals. It is essential to examine the performative effects and political narratives encompassing technological developments, and to emphasize their capacity to shape the world. This position reinforces the initial notion that interactions with AI and autonomous systems are redefining our relational and communicational frameworks. This position argues for a reflexive engagement with technology, where the imagination critiques and counterbalances the uniformity often propagated by mass media and hegemonic culture.

Later, through the lenses of Derrida's deconstruction, we observe that the technical reproducibility of a series of events alters the nature of the event itself, transforming it from its original form into one that is repeatedly experienced and thus changed. Reproduction and repetition occur through the technical re-inscription of events and the work of archiving. These manifestations occur in writing, photography, mass sound recording, television and cinema, as well as through words, listening and memory. These processes echo the complexity of the interaction between AI-generated content and human activity, highlighted at the start of the chapter. Technically produced objects can be reproduced without the presence of their original producer, outside the origin of production.

Building on these fundamental contributions, it is possible to propose further ideas for extending this view to posthuman communication. Therefore, integrating generative AI technologies into this post-representationalist framework opens avenues for rethinking the nature of domains such as creativity, intelligence and consciousness. In this context, generative AI becomes another form of posthuman expression, contributing to making sense of our world.

With generative AI, this fulfillment of the post-representationalist perspective for understanding posthuman communication suggests that we no longer consider language as a mere medium and representation of an external reality, but as a generative practice for creating and negotiating meanings within a post-human digital environment. Technologies based on generative AI modify the traditional notion that words and texts correspond directly to the world "as it is", advocating instead a perspective that recognizes language as an active participant in shaping our understanding of and engagement with the world around us.

2.7. References

Adorno, T.W. and Horkheimer, M. (1972). *Dialectic of Enlightenment*. Herder and Herder, New York.

Baudrillard, J. (1995). *The Gulf War Did Not Take Place*. Power Publications, Sydney.

Bijker, W.E., Hughes, T.P., Pinch, T. (eds) (1987). *The Social Construction of Technological Systems: New Directions in the Sociology and History of Technology*. MIT Press, Cambridge.

Boullier, D. (2016). *Sociologie du numérique*. Armand Colin, Paris.

Braidotti, R. (2013). *The Posthuman*. Polity Press, Cambridge.

Brown, A. (2019). Implications of generative AI in legal frameworks. *Journal of AI and Law*, 14(3), 205–219.

Critchley, S. (2005). Déconstruction et communication. Quelques remarques sur Derrida et Habermas. In *Derrida : la déconstruction*, Ramond, C. (ed.). PUF, Paris.

Derrida, J. (1967). *De la grammatologie*. Éditions de Minuit, Paris.

Derrida, J. (1978). *Writing and Difference*. University of Chicago Press, Chicago.

Derrida, J. and Prenowitz, E. (1995). Archive fever: A Freudian impression. *Diacritics*, 25(2), 9–63. doi: 10.2307/465144.

Derrida, J. and Stiegler, B. (2002). *Echographies of Television: Filmed Interviews*. Polity Press, Cambridge.

Doe, J. (2020). Linguistic models and AI: An emerging synergy. *Computational Linguistics*, 46(1), 159–178.

Giddens, A. (1990). *The Consequences of Modernity*. Stanford University Press, Redwood City.

Goldschmidt, M. (2003). *Jacques Derrida, une introduction*. La Découverte, Paris.

Habermas, J. (1981). *The Theory of Communicative Action*. Beacon Press, Boston.

Heidegger, M. (1958). La question de la technique. In *Essais et Conférences*, Préau, A. (trans.). Gallimard, Paris.

Heidegger, M. (1976). *Questions IV*. Gallimard, Paris.

Heidegger, M. (1983). *Lettre sur l'humanisme*. Aubier, Paris.

Horkheimer, M. and Adorno, T.W. (1947). *Dialectic of Enlightenment*. Stanford University Press, Redwood City.

Johnson, M. (2022). Generative AI and the future of human-machine interaction. *Technology and Human Behavior*, 37(4), 300–316.

Kittler, F. (1986). *Gramophone, Film, Typewriter*. Stanford University Press, Redwood City.

Lee, J. and Kim, Y. (2020). Innovative productivity: How generative AI influences creativity. *Innovation and Technology Journal*, 15(2), 117–134.

Lévi-Strauss, C. (1955). *Tristes Tropiques*. Plon, Paris.

Marcuse, H. (1968). *L'homme unidimensionnel*. Les Éditions de Minuit, Paris.

Martinez, R. (2021). Knowledge evolution in the age of AI. *AI & Society*, 36(1), 75–89.

Nguyen, L. (2020). Ethical considerations in the era of generative AI. *Ethics and Information Technology*, 22(3), 207–222.

Patel, D. and Singh, A. (2019). Generative AI: Bridging gaps between humans and machines. *AI Magazine*, 40(4), 34–45.

Rorty, R. (1979). *Philosophy and the Mirror of Nature*. Princeton University Press, Princeton.

Sadin, E. (2016). *La silicolonisation du monde : l'irrésistible expansion du libéralisme numérique*. L'Echappée, Paris.

Sadin, E. (2018). *L'intelligence artificielle ou l'enjeu du siècle : anatomie d'un antihumanisme radical*. L'Echappée, Paris.

Smith, J. and Jones, M. (2021). Recursive learning and AI: A new paradigm. *Journal of Machine Learning*, 12(3), 101–115.

3

Neuralink and the Birth of *Homo Silicus*

3.1. Introduction

Initially devoid of any particular aptitudes or skills, humankind only escaped the torments of a hostile environment thanks to their technical intelligence, which was dedicated, among other things, to artificially correcting their flawed nature, as highlighted by the Promethean myth.

On this basis, it can be argued that, unlike animals, which constantly adjust their needs and lifestyles to nature, humans, in order to survive[1], have had to adapt nature to their own. In this respect, humans are more *homo faber* than *homo sapiens*. Indeed, while they are not the only animal to use tools to ensure the conditions of their survival, they are, as Henri Bergson points out, the only animal to make "artificial objects" (by transforming existing materials or creating tools ex nihilo), to survive or to enjoy elements of comfort.

Even more interestingly, Bergson adds that humans are the only animal capable of making "tools to make tools, and of varying their manufacture indefinitely" (Bergson 2007, author's translation). So, by its very etymology, the term technique is perfectly suited to human, since it refers to an art of manufacture associated with logic, i.e. science, reasoning and discourse. All combined in a single signifier: the object.

Chapter written by Alexandre CARBONNEAU.

1 In fact, the black cockatoo, for example, uses natural materials to produce percussion sounds to attract nearby females.

Indeed, it is to the object – a term to be taken in its material, technical and scientific dimensions, as well as in its symbolic and analytical dimensions (as a reminder, object "a" is defined by Lacan as "the cause of desire"[2]) – that human devotes half their existence. The other half consists, more or less, of trying to understand why the objects thus created and/or possessed are so unsatisfactory.

Let us add that by bringing together the material and symbolic dimensions of the object, we arrive at the idea that the techno-scientific dimension of the material object serves primarily to rationalize the primitive and even archaic, fundamental and oh-so-structuring interest in the symbolic object, or object of desire forever lost.

In this light, the all-consuming passion for technological innovations and the practical achievements that flow from them seems not only logical, but also necessarily unsatisfying. Humans seem condemned to run after them, like greyhounds chasing a rabbit-shaped lure. An image which, in its formulation, ideally reproduces the Ouroboros approach to desire.

This futile race shifts the relationship between human and technology from its primary dimension (dedicated to the desire to survive) to its secondary dimension (dedicated to the desire to invent, reinvent and individualize ourselves – in short, to humanize ourselves).

The problem is that this second relationship engenders a consequence that becomes a trigger factor, namely the desire for power that humans exercise over their own nature – in order to satisfy the *desiderata* of the great "other" for analysts, of society for sociologists, of living together for philosophers, etc.

So much so that, in the end, in addition to surrendering their own nature to a ceaseless game of exploration/exploitation, the subject never ceases to project its desires, its individuality and even its idea of itself, into the system of technical objects. To put it another way, this system, based on the creation, consumption and constant rejection of more or less new objects, is

2 Psychiatrist Marc Darmon explains that, for Lacan, "object a is never the object of desire in the common sense. It is only a lure or a fantasy. Object a, in fact, never appears as such at the tip of the arrow of desire [...]. Object a is never in the position of desire's aim; it is either 'pre-subjective' or 'at the foundation of a subject's identification'". See Darmon (2014) with reference to Lacan (1973a, pp. 162–169, author's translation).

no longer just a stage, a link to desire: it becomes desire in itself, but not the cause of desire. An ideal position for making the quest for objects eternal, being eternally unsatisfying.

This is strangely reminiscent of the behavior of infants (children before language development), who use objects (all kinds of objects, including feces) to interact and, in a way, communicate with the world and, more specifically, with their mother. Furthermore, can we equate the shopping-addicted adult with the babbling baby? Not quite, because when it comes to this very human "greyhound race", the lure may not be the one we think it is. In other words, the lure is neither the object itself nor its possession.

Indeed, if technology promises to "arm its possessor with a power" (Russel 1947, p. 81–82, author's translation) that they initially lacked, it is also, by ricochet (as well as by a strange irony of fate), charged with a "connotation of the absolute" that induces in the contemporary subject a form of Hephaestus complex. By this, we mean the complex according to which a human is capable of molding their unsubmissive, imperfect and irrational nature into the form of a carapace worthy of the breastplate forged by Hephaestus for Achilles, i.e. indestructible and unalterable. After all, "we're worth it!". Able to manipulate the elements of nature and create new ones, humans consider themselves, by the same token, capable of forging their own individuality, their own mode of being, with nothing (or almost nothing) to hold them back.

But what precisely are the things that still hold them back? In reality, there's only one: death. Hence, the fierce determination to achieve *The Death of Death*[3].

While it is beyond the scope of this chapter to retrace the history of technology in detail, it is nevertheless appropriate to focus on the new relationship between humans and the technical and techno-scientific system they have created. It should be noted that, while human's innovations have enabled them to extricate themselves from their natural condition without ever totally losing control over the rationality of their technical system, this tendency has tended to wane since the end of the 20th century, under the influence of ever more ardent and passionate techno-scientific innovation.

---

3 Reference to Laurent Alexandre (2011).

Indeed, the refinement of Information and Communication Technologies (ICTs), based on the miniaturization of tools, the increase in their memory and computing power, as well as their convergence towards nanobiotechnologies (under the supreme aegis of cognitive sciences), are two phenomena that galvanize the most extravagant desires, at the same time as they exert on our consciences – including those of some scientists – a visceral, passionate, even mortifying attraction.

Why visceral? Because beyond mechanics, we are talking about guts, brains and intestines: those that we want to make indestructible to enable humankind to achieve immortality.

Why passionate? And why mortifying? Because passion defines, firstly, suffering: that of Christ. And secondly, according to the CNRTL, "the idea of excess, exaggeration, intensity" (author's translation). So, by jumping a few bridges, we could describe nanobiotechnologies as spheres of intense suffering. In other words, hell is here. We are there!

More calmly, we see innovation as a miraculous key that enables us to benefit from the "magical" power of technology, in the face of which its potentially harmful repercussions on our physical and psychic integrity leave us unmoved. Add to this the use of new terminology, often linked to the notion of addiction and dependency (cyberaddiction, fomophobia, etc.), and you have a model of pathologies characterized by the fact that sufferers do not want to get better.

In fact, something rather indeterminate, but close to an irrational sacralization of technology, is today accompanying the hope we are accustomed to placing in science. If this phenomenon can be clearly observed in the light of the aforementioned research into the euthanasia of death, a similar tendency pervades the entire living being, not only its body, but also its mind. So much so that technical and nanotechnological innovation is becoming the rallying sign for a new mythical order[4]: that of human endowed with intellectual and physical superpower.

4 If we choose to refer to myth, it is because these innovations are often presented on the basis of an idealized imaginary.

3.2. Artificial intelligence

While we readily applaud techno-medical developments, the technological revolution that is being added to them, involving AI, raises a number of questions. Indeed, by embodying the myth of a function transcending the individual in their most individualized possession, namely thought, AI absorbs the entire human imagination, transforming properly human activities (intellectual, creative, etc.) into purely functional ones. In other words, while many see AI as the elevation of technology to the level of human intelligence, it can also be seen as the reduction of the latter to the mere functional praxis of AI.

In so doing, beyond the results that AI enables us to achieve, it is the relationship we have with technology itself that is called into question. Thus, following Baudrillard (1968), we could say that "the fascination of AI is so great precisely because it is not that of a technical rationality" (author's translation)[5].

But is it scientific, psychological, philosophical or even creative? And, beyond that, does it have anything to do with the very notion of rationality? For several years now, philosophers such as Éric Sadin and Jean-Michel Besnier have been warning of the potential excesses of AI and, by extension, of practices that divert and distort rationality.

As far as we are concerned, we will simply point out that the deployment of AI raises the question of the erosion of our traditional relationship with technology. Without dwelling on the reasons for this erosion, it is important to determine how far it might extend. The answer is simple: the programmed obsolescence of our brains.

In response to this inevitable consequence of AI, some scientists are working on increasing our neuronal capacities. Even if, to this day, skepticism still prevails in the face of AI's predicted triumph over our brains, techno-scientific innovations are raising issues involving the hypothesis that our brains will adapt to the evolutions of this very AI. Witness the efforts of Chinese academics in the direction of cerebral eugenics.

---

5 To put it another way, and in keeping with the thinking of J. Baudrillard, AI fascinates because it remains, in the collective unconscious, that miraculous function capable of synthesizing all human efficiency in its strictly functional dimension.

At the same time, the interest of certain governments in enhancing our intellectual capacities is echoed in the laboratories of Silicon Valley, where billionaires, engineers and researchers are experimenting with electronic neuro-augmentation of these capacities. So much so that, alongside the development of a form of cerebral eugenics (in Asia), the specter of a *homo silicus* with intellectual skills augmented by cerebral implants is emerging (in the West).

Although still uncertain, this research is focused on the possibility of a human and machine fusion, using nanotechnological devices grafted into the brain. To date, these innovations have had little impact on social tranquility. However, their potential effects are a cause for concern, because, beyond representing an antidote to the obsolescence of our cerebral capacities in the face of the scope of AI capabilities, don't these innovations also run the risk of leading to a "normalization" of the human mind? More broadly, don't these innovations risk undermining the otherness that dictates and regulates the principles of collective desire, will and "power to live"?

With the aim of providing some answers to these questions, we propose the hypothesis that the human race is, in the more or less long-term, threatened by a "mathematical" artificialization of the living. To verify this hypothesis, we will analyze the risks posed by brain/machine hybridization on the mechanisms that condition mental ecology[6] and social ecology[7]. We emphasize that such hybridization cannot be achieved without deafness to a "collective intuitive knowledge"[8] that is part and parcel of our own

6 Faced with ecological disaster, philosopher and psychoanalyst Félix Guattari proposes a new way of thinking about things (Guattari 1989). Science philosopher Nicols Prignot explains that, for Guattari, "the current situation is the result of a way of thinking that doesn't know how to approach situations with the power and complexity they require. His 'ecosophical' thinking is not a massive solution, but a conceptual tool for comprehending situations through the use of three superimposed lenses: those of the ecology of nature, social ecology and mental ecology. This unprecedented kind of lens is intended to enable us to think about the conditions for producing human ways of life in a fragile world". He adds: "Ecosophy sees itself as an ethico-political articulation between three ecological registers: environment, social relations and subjectivity" (Prignot 2010, author's translation).

7 Mental ecology is concerned with the economic and social ecosystem that binds individuals together and influences individual behavior (see Guattari (1989)).

8 By the expression "intuitive knowledge", we refer to a powerful precognitive fund, common to the entire human species, which enables individuals to make the world.

humanity, as well as of our intrinsic and residual animality, which is also a constituent of our social cohesion.

3.3. Human intelligence as a multiple, subjective and non-reducible element

To deal with the mathematization of the living world, it is important to remember that human intelligence cannot be reduced solely to the notion of logical–mathematical intelligence and its valorization, including through the techno-sciences.

Indeed, far beyond "the single, general faculty measurable by the famous IQ" (Gardner 2008, author's translation), human intelligence also brings together a plurality of subjective aptitudes, resulting in what Howard Gardner calls "spatial, musical, kinesthetic, interpersonal, intrapersonal, naturalistic" and even "spiritual, existential or moral intelligence" (Gardner 2008, author's translation).

Yet, neuroscience[9], as does Elon Musk[10], favors a "purely objective approach to the thinking subject" (Falissard 2023, p. 136, author's translation). This approach, which relies on the inflation of technology (with medical imaging at the forefront), aims to uncover or enhance the mysteries of the human psyche, disregarding phenomena that cannot be empirically identified in the brain (i.e. everything that cannot be seen with medical imaging), and which nonetheless constitutes "the very experience of thinking" (ibid., author's translation).

This disregard for the subjective universe, and more generally for the complexity of the human psyche, is evident in the way these new sciences deduce the presence of mental disorders or behavioral problems from cognitive dysfunctions alone, and finds its aberrant culmination in the Neuralink project, based on the hybridization of human and machine.

In other words, this intuitive knowledge is based on a common, innate capital, and serves to interpret, understand and assume the place each of us occupies in what makes up the world.

9 Sciences on which transhumanist discourse is based.

10 As part of the Neuralink project, dedicated to the creation of a brain/machine interface.

Indeed, to say that the subject will increase their intellectual capacities simply because science is able to increase our memory potential by providing external electronic storage levers (or by implanting a search engine in our brains[11]) is tantamount to evading, through excessive objectivism, everything that shapes us – on a conscious and unconscious level – as thinking and speaking subjects.

In this respect, we need to overcome the current scientistic posture of abusively reducing the psychic and intellectual life of the individual to the sole mental manifestation immediately detectable in cerebral imagery during a cognitive activity performed by the subject themselves. But above all, the problem with this mathematization of thought is that it is based on the postulate of a match between the question asked (e.g. by the student connected to ChatGPT asking their machine to answer the question posed by their dissertation subject) and the answer given. To date, AI relies on a mixture of existing information, whether accurate or inaccurate. The human subject, on the other hand, can answer a question with another question. They can experience *lapsus*. They may not be able to answer. In short, it is in their failure to be perfect, or by their inventiveness and associations of ideas, that they signifies themselves as being human[12].

More broadly speaking, it is important to grasp that the reality of the human psyche is underpinned by antagonistic forces that elude any form of predictability, but which, moreover, opacify any immediate transparency in the subject. As mentioned above, we can associate with this last point the uncertainties, the gropings, the silences, as well as the contradictions and ambivalences that punctuate our discourse and actions, revealing the dissensual harmony[13] that sits between self and ourselves.

In the same way, we need to take into account the psychic value of language and what it outlines, and simultaneously betrays, in the stumbling blocks of speech such as a slip of the tongue, as well as in sudden stuttering, the use of words with double meanings, assonance or puns[14]. Nowhere are the subject's contradictions, inconstancies and inhibitions more stubbornly revealed than in speech.

11 Just as Raymond C. Kurzweil, the undisputed icon of transhumanism, would like.
12 This is why teachers prefer a copy containing errors and clumsiness, rather than one printed directly from the ChatGPT output.
13 Or to put it more simply, the coexistence of opposites.
14 For more on this subject, see Freud and Breuer (2002).

In this sense, the attack on the psyche, of which Elon Musk will soon be guilty, distances humans from an essential reference point: their unconscious. This shapes them through their own logic[15], as well as guarantees them otherness, everything that makes them a subjective singularity.

In order to identify the main issues at stake, we propose reflecting on what would happen to the process of autopoiesis[16] if humans, as such, were to plan to link their brains to an AI?

3.4. Autopoiesis versus the mathematization of life

While it remains tricky to pinpoint exactly what such an assault on the psyche will entail, the question nonetheless arises as to what effect this process will have, at the level of "self-production", given that realizing ourselves consists first and foremost of objectifying our psychic life and that which essentially conditions it.

Remember that realizing ourselves – existing (or "ex-sists", to use Lacanian terminology[17]) – implies first and foremost that we make the world, i.e. we define ourselves beyond (or beyond) the reality of a self isolated within itself. More generally, to produce ourselves requires that we go out to meet the other, and learn to be born with the other (and thus to "know" ourselves) in the vast lands of a self other than ourselves.

In this sense, self-realization is not limited to the rationalization of our psychic life by means of artificial arithmetical reasoning, devoid of all

15 Indeed, the unconscious is first and foremost a structure. Lacan states that it is "structured like a language" (Lacan 1966, pp. 856–864, tranlsated by Author). Here, he refers to his November 7, 1955 lecture, *"La chose freudienne"*, in which he touched on this idea already evoked by S. Freud. To put it more metaphorically, the unconscious is built like a tree whose branches, unique to each individual, vary according to the general psychic structure and are agitated by life's events.
16 Concept derived from the Greek auto (self) and poièsis (production, creation). In other words, *autopoiesis* is the ability to produce ourselves in interaction with others and the environment.
17 "It is not in every discourse that a saying comes to ex-sist" (Lacan 1973b, p. 467, author's translation) quoted in Leray (2014) Pascale Leray adds: "Indeed, not every discourse opens up, beyond the meaning conveyed by the word, to the real which would then give it another scope." This other scope can be summed up as the search, in the context of treatment, for the true word.

opacity and contradiction[18], if only because self-realization (the existential journey that falls to us) implies the coexistence of a psyche other than our own, also made up of chance, silences and constraints.

It goes without saying, then, that producing ourselves implies the expression of desire (the desire to make the world) as well as emotion, intuition, sensitivity and courage[19], the sources of human intelligence and its capacity for imagination, evaluation and transgression. It is, in fact, through such an assembly of specifically human qualities that humans can build themselves in the world and experience it among other humans and according to their own choices.

However, what would happen to these choices if they were relegated to the level of algorithmic logic, induced by nano-software implanted in our brains? It seems to us that this question is very close to our concerns. Given that, in parallel with the myth of increased intelligence, human/machine hybridization gives rise to another aspect that is strongly idealized by transhumanists[20]: that of the personalization of information and its alleged necessity.

In other words, we need to ask ourselves what will become of our choices in the light of this personalization and its supposed relevance to the reality of our individual and collective expectations. The question is worth asking, because what would happen if (or rather, when) the algorithms implanted in our bodies reduced the informational fabric, from which we try to understand the world, to pre-established alternatives deduced from our previous "navigations"[21]?

This question is significant, given that here and in this perspective, AI would not seek to increase the relevance of content by presenting information in its pure and immediately "prunable" form, information from

18 Or, to put it another way, by means of a surfeit of algorithmically articulated knowledge, aimed at understanding reality in the exclusive mode of a set of logically and rationally comprehensible elements.

19 Indeed, courage is a major component in the process of self-production, if only because it is a fundamental prerequisite for the self-abandonment necessary to be born to the other.

20 In particular, by Ray Kurzweil, chief engineer at Google. For Laurent Alexandre, the aim here is "to make the search engine the first artificial intelligence in history". See on this subject (Le Monde 2013).

21 We are deliberately using this commonly used term when we carry out searches using our current technological tools (PC, cell phone, touch-screen tablet, etc.).

which we would have chosen the "bad branches" to remove. Rather, AI would serve to orient the information given and, given that behind each AI is a company pursuing a commercial objective, to shape it based solely on the presumption of an individual expectation on our part, not strongly linked to some goods or services to be sold.

Information, as processed by algorithms implanted in our brains, would no longer serve to enrich our thinking, our imagination or our knowledge, but would be reduced to the mere fact of serving us. In line with the imaginary of "everything customizable", the algorithm would then, just as it does today, transform information into a "personal service"[22].

As subjects, then, making choices is tantamount to conforming to a single system of thought (namely, their own), based both on the concrete negation of otherness and on the abolition of all forms of conflictuality with others and the world. So much so that making a choice would no longer be a matter of choosing between two parties[23], or even of deciding between two or more ideas (according to our sensibility, emotions, etc.) or between antagonists: choosing would consist of aligning ourselves with alternatives mathematically pre-selected for us. The result would be the impossibility for each of us to deviate from the anticipated, one-dimensional and arbitrary algorithmic synthesis of our own choices.

In this sense, we would be obliged to take as personal choices, selections which, however, we would not have made by means of our psyche. What is more, we would find ourselves unable to appropriate, make, unmake and remake our own choices on the basis of the other, whose function it is to carve out for us "our own sense of the world and/or the world of our own sense" (Nancy 2007, p. 4, author's translation).

This is a key problem, for if, in the context of brain/machine hybridization machine hybridization, the increase in cerebral potential could enable the individual to adapt efficiently to the growing informational environment, this will not be possible without a redefinition of our perception of the world. To put it another way, rather than looking at the world in a contemplative and interactive (as well as ontogenetic[24] and

22 Jean Baudrillard wrote on this subject (Baudrillard 1968, pp. 252–253).
23 This is the dictionary definition of the verb to choose.
24 Relating to or specific to ontogenesis, the description of an organism's development from conception to death.

phylogenetic[25]) way, based on *koinè aisthésis*[26], don't we run the risk, in the end, of seeing it only in an exclusively algorithmic and egocentric way?

Hence, despite all the bewitchment of our increasing capacities, does this techno-scientific augmentation not risk reducing humans to the impossibility of projecting and then enriching the contours of their own existential frameworks, according to the mainspring they have used until now, namely the desire for transgression?

Remembering that the term "transgression" etymologically refers to the idea of going beyond, and that in this sense, transgression conditions the autopoietic dynamic in that it expresses the human desire to go beyond the perception of a world anemic to a rationally contingent order (cultural, technological and symbolic).

Producing ourselves as a unique subject implies that we not only know how to depart from established norms, but also how to conceptualize and appropriate new existential frameworks. Indeed, producing ourselves requires us to look beyond traditional norms and values for other ways of perceiving the world and other ways of organizing human experience (as artists and dreamers do), so that we can find, invent or reinvent our place in the world.

In light of the above, it can be asserted that increasing human capacities by mixing technological components with their typically opaque, contradictory and complex nature, thereby reducing their thickness and ambiguity, runs the risk that instead of machines becoming as intelligent as humans, we will give birth to humans devoid of humanity and poetry, thinking like machines, according to computer programming algorithms. These would be the elements of an irreversible regression, whereby human would present themselves in the guise of a resorption of intelligence, or, in the words of Jean-Michel Besnier, in the form of "an indefinitely manipulable abstraction" (Besnier 2006, p. 165, author's translation).

This mathematization of the living world must therefore be treated with caution, as it could affect our way of being, as well as, and above all, our

25 In relation to philogenetics, which designates the classification of living beings according to their inter-species and inter-species kinship links, from a hypothetical common ancestor.

26 In other words, on "common sense", "where all senses enter into conjunction with the senses of all" (Maffesoli 2012, p. 239, author's translation).

desire for transgression. It would exhaust the pleasure of seeking and building knowledge capable of emancipating us from immediately perceptible realities. In this way, mathematization could destroy our sensitivity, our imagination and our curiosity.

It is this curiosity that gives rise to the desire to learn about our environment so as to emancipate ourselves from finalized knowledge. Curiosity, far from assigning us to learning a series of academically validated truths, enables us to comprehend each truth for ourselves, sometimes even beyond socially pre-established determinations.

More broadly, curiosity offers each individual the means to divide knowledge into as many parcels as sensible reason has principles of reality. For this reason, such a reduction of being and its unpredictable potentialities can only lead to the irreversible submission of its desires to the arithmetical schema of finalized knowledge alone[27].

3.5. AI: a new aspect of heterotelia[28]

Discussed earlier, F. Guattari's notion of ecosophy is an original approach to thinking about the environment, based on three elements: mental ecology, social ecology and the ecology of nature.

The first two elements, thought of by the human subject, are marked by the encounter of particular moods and affects. However, with the mathematization of the living world and human thought, this mental ecology – and, beyond that, the way we think about the environment and the relationship between humans and the environment – runs the risk of being deprived of the "singular" relationship (in both senses of the term) that each of us has with the world.

It is this regressive scenario that is being written in the obsession with brain/machine hybridization. It is a scenario on the verge of becoming reality, and one that brings with it a dehumanizing force that threatens the living power of emotions. Emotions that are told, exchanged, shared and reinvented through storytelling, with all its variations and stumbling blocks. To be convinced of this, we need only observe the culpable inconsistency of

27 In other words, knowledge assigned to a functionalist, industrialist purpose.

28 Heterotelia characterizes a result, outcome and consequences other than what the author had in mind and intended as an objective.

the techno-messianic myth of the superhuman, driven by the transhumanist discourse that contravenes the very root of human intelligence, by instructing science to purge it of its intuitive and emotional dimension.

But what is an intelligence without intuitive and emotional strength? An intelligence condemned to no longer follow the tree-like itinerary of possibilities that our intuitions sketch out for us on "mental maps"[29]? The answer is simple: it is an inaudible intelligence, excluded from any dialogue.

The AI with which we will be endowed by means of human/machine hybridization will therefore remain a decision-making body. But what else could we do with it, and what else could we expect from it? Beyond that, what will become of our intelligence without the presence of typically human affects, capable of grasping and sublimating words, images, events and sounds in a reticular flow of concepts and meanings?

These questions need to be asked, because let us not forget that intelligence starts from the body and is then deduced in questions and answers. It is an extrapolation, through problematization, of the meaning already intuitively explored by our sensibility, otherwise known as intuitive precognitive knowledge[30]. It is this intuitive precognitive knowledge that precipitates words, while words run towards their own meaning. This intuitive knowledge marks and gives life to conceptual infinity. Nietzsche said as much in this unequivocal sentence: "Of all the varieties of intelligence so far discovered, instinct is the most intelligent" (Nietzsche 1886).

3.6. Functionalism as the sole object of intellectual accomplishment

Once again[31], the transhumanist discourse is surrounded by a heterotopic ideal induced here by the myth of an intelligence purged of its emotions. The concept of human/machine hybridization neatly encapsulates this myth,

29 To take up a notion amply developed by Tony Buzan, its promoter.

30 We borrow this notion from Gorz (2003), 1978 expanded edition published by Le Seuil, Points collection, which adds the text *"Écologie et liberté"* (1977).

31 Our previous research has already enabled us to face up to this observation, through a philosophical analysis dedicated to the transhumanists' ultimate project: the euthanasia of death.

making affect, sensitivity, intuition and transgression the pejorative expressions of an intelligence without "operative value".

However, the desire to exhaust the meaning of intelligence at all costs in an officially operative function, so that this same intelligence can never be identified with any function other than the one previously assigned to it, leads us to reduce the very nature of our psychic and psychophysiological qualities to the single principle of techno-scientific functionalism.

In fact, by making functionalism the sole criterion for the legitimacy of intelligence, and the sole object of its accomplishment, this same intelligence runs the risk of silencing any other intellectual manifestation that does not follow the intentional logic of "practical instrumentality"[32]. The result would be the emergence of an all-powerful, expert intelligence, subject to the arbitrary, practical instructions of a technomental system that seals its sole skill in automated problem-solving, following the logic of adapted algorithmic processing.

What we call intelligence would then become a machine destined to immerse our minds in the emptiness of functional utopia. What is more, we might add that, reduced to the technobiological workings of cold but infallible reflection, intelligence runs the risk of imposing itself on the mind as the obsessive alibi for a "narcissistic manipulation" (Baudrillard 1968, p. 145, author's translation) of our practical and abstract knowledge. To put it more simply, human intelligence, rather than "shining its light a little further than itself, towards the other, over there, etc." [33](author's translation), would work solely towards itself, according to the involutionary rule of functional and performative differentiation.

Indeed, freed from its inclinations towards sensitivity and above all reclined in its connotation of infallibility and omniscience, intelligence would then deploy its determinations – both subjective and objective – against the other[34] (or, if not against, at least without the other). Put another way, it would have no other direction than that of a strictly narcissistic

32 Terms borrowed from Jean Baudrillard (Baudrillard 1968, p. 165, author's translation).

33 Here, we borrow part of Christian Bobin's definition of intelligence: "Intelligence is the solitary strength to extract from the chaos of one's own life the handful of light sufficient to illuminate a little further than oneself, towards the other out there, like us, lost in the dark" (Bobin 1996).

34 The term "against" is meant here in a competitive logic.

fascination with making humans appear, in their own eyes, as the most miraculous tool of their techno-scientific system.

Jean Baudrillard wrote: “We are more sensitive to the disruption of human relations by the absurd and totalitarian intervention of technique, we are less sensitive to the disruption of technical evolution by the absurd and totalitarian intervention of the human” (Baudrillard 1968, p. 164, author’s translation). Yet, it is this second point that attracts our attention in the project of human/machine hybridization.

Indeed, the techno-scientific prophecies of omniscience pose a worrisome ideal that intrudes into the very intimacy of thought, and which will soon impose on humankind full and complete adherence to the regressive dream of instrumental functionality. All the more so as, despite the expansion of our intellectual coefficient, we will lose to this subservient intelligence everything that shapes our differential coefficient.

But let us not make a mistake: such a rationalization of intellectual ends will not only result in the establishment of an exclusive regime based on practical instrumentality. Nor will it reduce its effects to a strictly narcissistic fascination. In fact, at the same time as it opposes any new way of feeling and expressing what we are, this rationalization of intellectual ends will close humans and their intelligence “to the relationships of meaning that are expressed by language and signs, but which cannot be reduced to the latter”[35] (author’s translation).

In short, such rationality would distinguish being to the existence of a meaning expressed beyond words, “in the event-related dimension of being” (author’s translation) [36]: that dimension which is not yet there, but which is to be made. Echoing Bakhtin, Maurizio Lazzarato speaks of “a position internally out of” (author’s translation)[37], meaning out of a world already there, already communicated.

This is where we see the real danger of human/machine hybridization. In the sense that such a redefinition of intelligence and its determinations would paralyze any possibility of a human singling themselves out, as well as of

35 Here, we take up an idea developed by this philosopher (Lazzarato 2004, p. 192).
36 Ibid., p. 195.
37 Ibid.

subjectivizing themselves beyond a shrunken dialectical system[38] reduced to the sole signifiers valorized by the algorithmic power of the computing machine. A system whose efficiency means that we no longer even try (or indeed, could we?) to envisage alternatives other than those imposed on our minds by the dominant modes and models of communication.

In short, we would find ourselves in an autarkic, inward-looking world, in which no one would take part in the making, organizing and becoming of the world, except within the alternatives, undulations and bifurcations determined in advance, to meet the economic, political and techno-scientific imperatives of a future era.

From there, if we go beyond the strictly jubilant aspect of the brain/machine interface project, we note that the prospects of such "omniscience" induce, in many respects, the possibility of a radical inversion of their own finality. For all that, this heterotopic reversal[39], this "paradox of consequences" as Marx would have called it, disappears from transhumanist thinking and from Elon Musk's reflections, behind the shameless admission of a "crime against the real world" (Baudrillard 2000, p. 65, author's translation).

3.7. Conclusion

Throughout this chapter, we have attempted to describe the sometimes zealous contribution of scientists, engineers and billionaires to the disastrous enterprise of human/machine hybridization. To do so, we have taken a critical look, without seeking to systematize our disapproval of techno-scientific and medical innovations. We note that such an approach is necessarily necessary, given that, whatever the fables on which the myth of the augmented human is based, who would oppose the deployment of cerebral implants capable of repairing decaying brains, trapped by pathologies such as Alzheimer's or Parkinson's?

This being the case, we felt it important to point out that, at a time when techno-science is prophesying the triumph of omniscience, we are also

38 In other words, a system that allows for singular discussion, reasoning, interpretation and questioning of reality, of ourselves and of others.

39 The notion of heterotelia was conceived by Jules Monnerot, and derives its meaning from the Greek words *eteros* (other) and *telos* (end). Etymologically, heterotelia refers to the other end.

witnessing the wholesale execution of our ways of making, unmaking and remaking the world, according to the rules of unpredictability and indeterminacy.

More precisely, transhumanist thinking, and that of Elon Musk in particular, in a regressive drive towards the best[40], seeks to mathematize an imaginary without compensation, thus forcing the marginal laws of human intelligence to take refuge, either in the algorithm, or in madness (a bug). In this respect, the astonishing coexistence of randomness and arithmetic raises the question of the future of undecidability.

In anticipation of this, we sought to problematize the phenomenon, questioning the philosophical, psychological and sociological consequences potentially linked to taking such a path. In doing so, we noted that once assigned to the order of algorithmic calculation, our choices, our sensibilities, as well as our desires and ways of making the world, would become disassociated from an essential component: their differential coefficient.

But is it not this same coefficient that, at every turn, clears our minds of the absurd human pretension to believe that we have to be anything other than what we already are: An end in itself? And is it not this same differential coefficient that, in so many ways, pits our creative forces against the temptations of uniformity?

Beyond its subjectivizing action, the differential coefficient represents for humans a consolation, the last parapet against their own despair. To take it away by subordinating the potentialities of this same differential coefficient to calculation is to wrest from us the guarantee of our freedom in relation to ourselves and others. In other words, it is to deprive us of the wings which, at every moment, bring us back towards a dizzying horizon: our reason for existing.

3.8. References

Alexandre, L. (2011). *La mort de la mort*. JC Lattès, Paris.

Baudrillard, J. (1968). *Le système des objets*. Gallimard, Paris.

Baudrillard, J. (1970). *La société de consommation*. Denoël, Paris.

40 As the enemy of good, as Montesquieu would have pointed out.

Baudrillard, J. (2000). *Mots de passe*. Biblio Essai, Paris.

Bergson, H. (2007). *L'évolution créatrice*. PUF, Paris.

Besnier, J.-M. (2006). *La croisée des sciences, questions d'un philosophe*. Le Seuil, Paris.

Bobin, C. (1996). *L'inespérée*. Folio, Paris.

Darmon, M. (2014). De l'objet perdu à l'objet trou... *La revue lacanienne*, 15, 29–32. doi: 10.3917/lrl.132.0029.

Falissard, B. (2023). Troubles du neuro-développement : pourquoi "neuro" ?. *Enfance et Psy*, 97, 29–34. doi: 10.3917/ep.097.0029.

Freud, S. and Breuer, J. (2002). *Études sur l'hystérie*. PUF, Paris.

Gardner, H. (2008). *Les intelligences multiples, la théorie qui bouleverse nos idées reçues*, Bonin Y. (trans.). Retz, Paris.

Gorz, A. (2003). *L'immatériel, valeur, connaissance et capital*. Galilée, Paris.

Guattari, F. (1989). *Les trois écologies*. Galilée, Paris.

Lacan, J. (1966). La science et la vérité. In *Écrits*. Le Seuil, Paris.

Lacan, J. (1973a). *Le Séminaire "Les quatre concepts fondamentaux de la psychanalyse"*. Le Seuil, Paris.

Lacan, J. (1973b). L'étourdit. In *Autres écrits*. Le Seuil, Paris.

Lazzarato, M. (2004). *Les révolutions du capitalisme*. Les Empêcheurs de penser en rond, Paris.

Le Monde (2013). Google et les transhumanistes. *Le Monde.fr*, 18 April [Online]. Available at: https://www.lemonde.fr/sciences/article/2013/04/18/google-et-les-transhumanistes_3162104_1650684.html [Accessed 14 June 2024].

Leray, P. (2014). Au-delà de la parole : le dire rappelé à l'ex-sistence. *L'en-je lacanien*, 23, 43–57. doi: 10.3917/enje.023.0043.

Maffesoli, M. (2012). *Homo eroticus. Des communions émotionnelles*. CNRS Éditions, Paris.

Nancy, J.-L. (2007). L'être-avec de l'être-là. *Cahiers philosophiques*, 111(3), 66–78. doi: 10.3917/caph.111.0066.

Nietzsche, F. (1886). *Par-delà le bien et le mal (Jenseits von Gut und Böse – Vorspiel einer Philosophie der Zukunft)*, Bianquis, G. (trans.). Éditions Montaigne, Aubier.

Prignot, N. (2010). Retour sur les trois écologies de Félix Guattari. *Philosophie de la Nature* [Online]. Available at: https://etopia.be/blog/2010/06/01/05-retour-sur-les-trois-ecologies-de-felix-guattari/ [Accessed 28 June 2024].

Renucci, F. (1970). L'autre n'est pas une donnée, altérités, corps et artefacts. *Hermès La Revue*, Paris.

Russel, B. (1947). *L'esprit scientifique et la science dans le monde moderne.* Éditions J.B. Janin, Paris.

Part 2

Questioning Ethics at the Heart of AI

4

The Influence of AI on Love: The Role of Dating Sites in Romantic Encounters

4.1. Introduction

Love, that strong feeling of affection and attachment, often leading the person who experiences it to adopt a particular behavior and, if this love is shared, to end up in a loving relationship. As such, love is a major factor in social relationships.

This is why love is as much about social norms and values as it is about people's feelings. Immersed in a hyper-connected space, our social interactions, including our romantic relationships, are stimulated and challenged by the evolution of social networks and digital platforms, which push us to meet a multitude of people who are all potentially "our soulmate 2.0".

Our individualistic society encourages us to make a series of consumer choices when looking for and finding our soulmate, based on finding the profile of the person who matches our expectations. This approach to romantic partnership is likely to call into question the model of romantic love that prevailed as the reference model for love relationships; when it relies on online dating, which is the practice of using dating sites to find a partner, it modifies the conventional dating process that traditionally led to the amorous encounter.

Chapter written by André BOYER and Faranak FARZANEH.

Indeed, chance plays less of a role in online dating than in face-to-face amorous encounters, in the course of the current lives of potential partners, whether these encounters are partly chance or not (Finkel 2012). With online dating, the original face-to-face encounter and its fortuitous nature fade away.

To analyze the effects of digital technology on the transformation of the love relationship process and, ultimately, on the perception of love it engenders, we devote the first section to the evolution of human relationships through social networks (section 4.2), before presenting the effects of dating sites on modern love (section 4.3) and analyzing them through a theory of love (section 4.4).

4.2. The evolution of human relations through social networks

4.2.1. *A quasi-permanent interconnection*

Over the past 20 years or so, connected devices have become widely democratized: desktop and laptop computers, touch-screen tablets, smartphones, games consoles and TV sets. These devices have led to a near-permanent interconnection between individuals, with consequences for the nature of their relationships.

The Internet has revolutionized a number of social spheres, including communication, politics, sociability, identity, cultural creation, human work and the economy. In this chapter, we focus on its effects on interpersonal and, more specifically, amorous communication, in the particular case of dating platforms.

Email, SMS and chat are just some of the communication devices that are expanding the possibilities of interpersonal communication. This diversification is leading to communicational inflation, which is clearly transforming social relationships. Indeed, the convergence of communication and information tools, as well as the portable nature of new communication terminals, have modified users' communicative contexts. These changes have consequences for the way contacts are organized.

With the Internet, individuals are hidden behind their screens, giving a new place to the body. The stakes for identity are high: online life (OL) and in real life (IRL) intertwine to the rhythm of the uses made of these objects.

Individuals can now easily multiply their contacts. As a result, the Internet has made socializing more complex, making it easier to connect with people close to you, as well as with strangers, thanks to the disembodiment of relationships.

Yet the ways in which social relations are built, managed and maintained are linked to the tool that mediates the relationship. This is why the new connected communication devices are vectors of new ways of creating, managing and maintaining links, and in particular, the continuous connection that enables these new tools to offer a 24/7 communication space.

These new communication tools have led to the creation of both public and open social networks, as well as closed social networks, including dating sites.

4.2.2. *The significance of closed social networks*

The growth of social networks has propelled them into widespread use, reaching 5 billion users in 2024, or 62.3% of the world's population. Social networks, whether open or closed, offer opportunities for individual and collective expression at the individual level, contributing to the formation of personal identity, while providing emotional support in the form of comments, likes, shares and reactions.

According to Tronca and Sita (2019), social networks can be seen as structures of interaction between a set of actors (individual, institutional, etc.) who are connected to each other by links that enable a flow of resources (information, support, material help, belonging).

Closed social networks, on the other hand, emphasize private interaction patterns, favoring relationships between individuals and groups of friends. This model offers users protection from external threats, and immediate recognition within their group and social status. Closed social networks followed the emergence of the Internet, with Web 2.0, in which users become actors by contributing content to websites such as blogs. They enable virtual communities to be formed, bringing together a group of people who interact online. They foster social relations between users by offering interactive platforms where users can connect, communicate and exchange their day-to-day experiences, and in these communities –

pioneered by Facebook – the physical face of the communication process is eliminated.

They therefore encourage the maintenance of community ties, friendships and romantic relationships (Dang Nguyen and Lethiais 2016), as well as offer the opportunity to meet new individuals. These two complementary, if not opposing, processes reveal the complexity of online identity, where individuals are free to constantly create and reinvent their presence on social networks.

New types of relations are also emerging, compatible with the freedom of expression of each individual, where people aspire to flexible, non-binding relations, identifying with a multitude of attributes, the maintaining weak relations, enabling the creation of new relationships and favoring the tendency to associate with similar individuals.

The world of gaming can be associated with closed social networks, providing a microcosm of contemporary social dynamics. While video games have long been regarded as mere entertainment, their meteoric rise as a cultural industry and a space for socialization needs to be analyzed as a societal issue. Indeed, games have given rise to virtual communities where individuals share a common passion for interactive worlds.

The links forged by social networks carry risks for users. Indeed, by facilitating unprecedented connections and sharing, social networks expose users to risks that pertain to the veracity of exchanges. Threats such as identity theft, piracy and the dissemination of false information are just a few examples of the dangers that lurk online. Piracy represents a significant risk for social network users, characterized by unauthorized intrusions into personal accounts, with the intention of stealing sensitive information or perpetrating fraudulent acts. Identity theft is one such risk associated with the use of social networks: personal data, such as dates of birth, addresses and various numbers, can be used by malicious actors to impersonate a person.

The consequences are the spread of false information, which is an omnipresent risk on social networks, for example, in the form of deep fakes, videos or audio recordings manipulated by artificial intelligence to make people say or do things they have never said or done.

All the characteristics of social networks apply to dating sites, which are their culmination in the sense that the creation of social relationships is their explicit objective.

4.3. Modern love in the context of dating sites

Love is a polymorphic concept, since it can take many forms, such as human love, divine love, paternal love or maternal love, with behaviors that differ according to the object of love. The history of each individual is that of the construction of their love model and the behaviors associated with it.

Love encompasses a wide range of emotions, behaviors and attitudes. It is often considered a deeply personal and universal experience, but it can vary considerably according to cultural contexts, individual experiences and interpersonal relationships. Some dimensions of love can be classified as follows:

– Romantic love, encapsulates three types of passion, with a strong emotional and physical attraction to another person, a dimension of intimacy in the form of a deep connection and mutual understanding, as well as commitment, i.e. a willingness to maintain the relationship and make efforts to overcome challenges together.

– Family love involves affection and support between family members, manifested in care, attention and mutual support, as well as a sense of responsibility inspiring a sense of duty and accountability to family members.

– Platonic love is founded on complicity constituted by a strong connection based on common interests, shared values and camaraderie, and complemented by a sense of loyalty, implying mutual support and faithfulness in good times and bad.

– Self-love involves accepting our own qualities, as well as recognizing and working on our shortcomings, and taking care of ourselves physically, emotionally and mentally.

– Universal love is associated with altruistic actions and a desire to contribute to the well-being of others, with a willingness to treat all human beings with fairness and dignity.

– Spiritual love is based on a transcendent connection with a higher power, the universe or the divine, often associated with religious or spiritual practices, and an inner peace that corresponds to a deep sense of peace, contentment and understanding of ourselves and the world.

In all cases, love is a powerful force that profoundly influences human behavior and relationships, because it is expressed as an emotion, a choice and a series of actions aimed at creating and maintaining meaningful connections with others and with ourselves. In this chapter, we focus on the first two meanings of the term, namely love-passion and family love, which are particularly affected by the development of dating platforms.

In this context, children are born with a repertoire of attachment behaviors designed to call upon or maintain their parent. Initially, these behaviors are expressed spontaneously, then, depending on the adult's responses, they are modulated and behavioral strategies are developed that seem effective in obtaining the closeness the child is looking for. In this way, a child builds themselves up by following a model relating to their own needs, linked to the model of their relationships with those around them. In relation to these two models, they form expectations about how their relationships will unfold.

Our first perception of love therefore depends on the one instilled in us by our parents from a very early age, depending on how they behaved with us and with each other (Miljkovitch and Cohin 2007). These behaviors define our first conception of love, which evolves as a function of our successive experiences. These experiences generally lead the individual to form a couple, which, unlike the parent–child relationship which is always imposed on them and whose evolution they undergo as a result of society's own evolution, can be voluntary (Goffman 2002).

4.3.1. *The evolution of the couple*

Unlike the period when the couple was instituted within the framework of a marital contract that linked conjugality, filiation and parenthood in an endless life together, today's couple in the Western world is held together only by the feeling that unites its members (Neyrand 2015). What is more, according to Neyrand (2004), society has moved on from the symbiotic couple model, with the wife staying at home, to the "duo" couple model, which preserves the autonomy of both partners, each with their own activities that are not necessarily complementary.

However, the evolution of couples has been influenced by socio-economic, cultural and technological factors, including gender equality, with women playing a more active role in the job market and family decision-making. This is also due to the diversity of family models, which include not only married heterosexual couples, but also same-sex couples, single-parent families and blended families, and the acceptance of cohabitation without marriage, with many couples choosing to live together before getting married, or with no intention of marrying.

As a result, marriage rates have fallen in many countries, while divorce rates have risen, particularly in Western countries, and marriages tend to occur later in life, often after couples have achieved a degree of professional and financial stability. These behaviors are often linked to economic aspects, with two often professionally active partners sharing financial and domestic responsibilities, and economic pressures, such as the high cost of living and job instability, which can affect the stability of relationships.

This shift in the model of the couple that began in the 1970s, with the rise of values around personal freedom and self-fulfillment, as well as equality of roles within the family, is now becoming the norm in interactions between men and women. Today, the primary function of the couple, before considering procreation, is to contribute to the fulfillment of both partners (Neyrand 2015). The couple becomes a space for interactions geared towards affective, sexual and social gratification, with a view to reciprocal self-realization through the other.

Dating sites, born of ubiquitous social networks, are part of this new social framework for couples.

While communication technologies (social networks, instant messaging, video calls) have changed the dynamics of social relationships, enabling constant interaction, dating sites have transformed the way people meet and form relationships, exerting an influence on the way couples meet.

4.3.2. *Exchanges on dating platforms*

The role of the Internet is part of the preceding social context. The evolution of digital technologies and dating platforms has profoundly transformed the way individuals interact and form romantic relationships. Far from being mere dating facilitators, these tools influence users' perceptions of love, relationship expectations and behaviors. Dating sites are designed to facilitate exchanges and encounters for couples looking to get together.

Indeed, since electronic communication makes it possible to exchange throughout the day with our partner wherever they may be physically, it allows each person to free themselves from any social constraint allowing couples to become desynchronized (Le Douarin 2002; Demonceaux 2014), as shown by the example of the use in bed, and as a couple, of smartphones with headphones to exchange with others (Salmela et al. 2019).

As a result of these exchange possibilities, the development of dating sites directly challenges the notion of love. The use of dating sites is leading to a change in human behavior towards couples, and is thus helping to modify the concept of love. It is also changing the probability of encounters between people from different social classes, occupations or locations (Duteille 2007).

While the majority of encounters still take place outside the Internet (Bergström 2021), platforms have seen their relative importance grow, due to the increasing ease of Internet access thanks to cell phones and the commercial methods of digital matchmaking services.

These relationships have been perfected by allowing users to create their own online profiles, describing the characteristics they are looking for in their desired partner, and uploading a set of photos to their profiles. These dating sites, in turn, allow users to search other users' profiles using a search engine that allows them to filter by age, location, religion and other variables. In addition, applications installed on smartphones can use location data to adapt their recommendations to the context (Marquet 2009).

The dating environment created by online sites modifies people's relational behavior. When meeting at work, at university or at a party with friends, for example, the protagonists are careful not to reveal their desires and feelings too quickly, making the approach more gradual to ensure that expectations are mutual. Online, since the stakes are explicit, the seduction time is shortened and the transition to sexuality accelerated. This acceleration also seems to characterize the later stages of pairing up: a study from the USA indicates that couples who met online marry faster than other couples (Rosenfeld 2017).

If virtual encounters speed up the process of getting together, they also change the parameters that form the basis of the encounter, so that getting close to someone for sexual interaction is no longer taboo – a step that would probably have been more difficult to take before virtual encounters.

4.3.3. *The online dating market*

The number of people using these sites is constantly growing, with 370 million active users worldwide by the end of 2021, according to Statista Digital Market Outlook (2018), while the range of choices on offer is similarly expanding. In France, it would appear from a series of surveys that almost 25% of people find a partner through a dating site.

The data collected by these sites enables common points of interest to be highlighted, such as the type of sexuality, the relationship sought and the number of people involved. The latter can then focus on more specialized themes such as the users' religion or even their eating habits.

In all cases, dating sites put forward the prospect of complementary partners based on a rigid conception of the amorous feeling (Pagotto 2016), due to the sites' purpose of bringing together beings sharing commonalities.

It is therefore a market (Marquet 2009) in which the users of these platforms seek to buy and sell themselves, as a product. This approach to the other, to the couple and to love, has effects on people's behavior during and after the exchange on the market, which is not immune to the risk of falsifying expectations generated by competition between the different members of these platforms. Also, once the exchange has been established, partners are likely to conduct their relationship in a different way from those who met in the traditional manner.

As users' interactions on these dating sites are linked to the competition that prevails (Hertlein and Webster 2008; Marquet 2009; Demonceaux 2014), dating sites allow the user to practice a multi-partnership (Marquet 2009), thanks to the confidentiality that the sites ensure, thus undermining one of the recurring concepts of the love relationship dynamic, namely monogamy (Richter 2014).

Furthermore, online dating can make people feel interchangeable, a feeling reinforced by the brevity of interactions, which undermines their commitment. This commodification of love therefore clashes with the ideals surrounding love and its passionate components (Guéret 2020), notably the exclusivity of the love relationship with a partner. Factors in favor of partner rotation are, moreover, integrated by the sites' communication (Bergström 2019), since the multiplication of partners is justified to participants by the need to make the right choice and is facilitated by the great discretion of the online dating concept (Marquet 2009; Bergström 2019), which allows a high degree of autonomy in the way the user's private, sentimental and sexual life is perceived, in contrast to conventional encounters such as a party organized by friends, a meeting in a bar or at work, which are subject to the gaze of other people present in these various environments.

The question then arises of comparing the nature of couples created via dating sites with couples who met conventionally. Simplifying, the former can be linked to a short-term conception of the couple based on the satisfaction of immediate pleasure, while the latter are attached to a more traditional, long-term conception of a love based on stability and on traditions anchored generally through parents and societal norms (Pagotto 2016).

However, we cannot ascribe users' tendency to be satisfied with ephemeral relationships determined exclusively by this proliferation of choice, as we can also argue that dating sites are merely the consequence of an expectation that can finally be expressed. In other words, are online matchmaking sites adapting to social change, or conversely, are they instigating it? At the very least, they are facilitators, and they are influencing the nature of love encounters.

Nevertheless the consumerist logic of the market introduces a conflict of values into online dating experiences. Users are looking for love while navigating a consumer space where immediate gratification and infinite choice can lead to a codestruction of value (Illouz 2007; Minina et al. 2022), which leads us to analyze the influence of dating sites on the love relationship dynamic.

4.4. Analysis of the effects of dating sites on romantic relationships

This influence can only be multidimensional, given the number of variables involved. To try to grasp its contours, we have drawn on the role of dating sites on romantic relationships (section 4.4.1), confronted with a theory of love (section 4.4.2).

4.4.1. *The effects of digital platforms on the dimensions of relationships*

Online dating platforms have a significant impact on the nature of romantic relationships through several dimensions: the diversity of choices, the ease of dating, the opportunity to establish new types of relationships, the possibility of selecting a partner on the basis of specific criteria, the evolution of expectations and the modification of interactions between partners:

– Online platforms provide greater access to potential partners based on the diversity of choice, as users have access to a vast database of profiles, increasing the chances of meeting someone with similar interests or values. What's more, dating is no longer limited to a specific geographic region, enabling individuals to connect with potential partners from all over the world.

– They act as dating facilitators, enabling people to meet faster than conventional dating, with matching systems based on algorithms, especially as online conversations allow site users to get to know each other before meeting in person, reducing the stress of first-time encounters.

– They also provide the opportunity to offer new types of relationships, whether short-term or long-term. In the first case, easy access may encourage them, but dating sites also offer people the opportunity to find serious partners and build lasting relationships.

– They enable participants to select a partner on the basis of specific criteria, using filters and preferences: users can specify detailed search criteria, which can guide choices and influence initial compatibility.

– They have a superficial aspect, as focusing on profiles and photos can make dating more superficial and appearance-based.

– They allow expectations to evolve, due to the pressure of perfection: the abundance of choice can create pressure to find the "perfect" partner, which can be unrealistic and lead to disappointment.

– They change commitment dynamics, with some people taking a more detached approach to relationships due to the ease of meeting other potential partners.

– They have an impact on communication and interaction, with the prevalence of instant messaging and video communication changing the way people get to know each other.

– They make it difficult to judge a person's authenticity and intentions online, which can lead to misunderstanding and disappointment.

Dating platforms have transformed the way people meet and maintain relationships, offering both increased opportunities and new challenges, which can be set against an outline of the theory of love.

4.4.2. *The effects of dating platforms through a theory of love*

The nature of love is considered a universal experience strongly associated with sexual arousal, and capable of having a strong effect on a person's emotional, cognitive and behavioral dimensions (Feybesse 2015). Analyzing this nature of love, Sternberg (1986) proposes to distinguish three components in love: intimacy, passion and commitment.

Intimacy encompasses the feelings of closeness, connection and bonding experienced in loving relationships. The degree of intimacy depends on the degree to which emotions, brakes, motivations and attitudes are shared:

– Passion is first and foremost the attraction between two individuals, encompassing both sexual consummation and romance. For example, the excitement of reuniting with our partner is passion.

– Decision or commitment encompasses two decisions over two different durations:

- in the short-term, the decision to love the other person;
- in the long-term, the commitment to maintain that love.

In an attempt to understand why some relationships tend to end more quickly than others, Sternberg (1986) identifies time as a moderating variable, postulating that it brings about fundamental changes to the nature of the love relationship dynamic. The three dimensions pertaining to relationships – intimacy, passion and commitment – are linked to the duration of the relationship.

The effects of digital platforms on the dimensions of the love relationship dynamic can be summarized as follows, each of which has its own impact on the duration of the relationship:

– diversity of choice: access to a vast database of profiles;

– easier dating: faster meetings thanks to algorithm-based matching systems;

– new types of relationships: possibility of finding both serious partners and short-term relationships;

– partner selection: specific criteria via filters and preferences;

– changing expectations: pressure for perfection and risk of disappointment;

– changing commitment dynamics: more detached approach to relationships;

– difficulty in judging authenticity: risk of misunderstanding or disappointment.

By confronting the effects of digital platforms with Sternberg's theory of love, it appears that these platforms can have a negative impact on commitment, as users seek love while navigating a consumer space where immediate gratification and infinite choice can lead to a codestruction of value.

Our assumption here is that commitment, or the third dimension of the relationship dynamic directly linked to time, leads to a marital purpose in the traditional definition of love. However, on the one hand, the individualistic vision of society implies the extension of sexual freedom, which is not conductive to long-term commitment (Bankov 2019). On the other hand, the digital age brings tools to implement, in a practical way, this extension of sexual freedom, to call into question commitment as analyzed by Sternberg (1986), the first two structural changes, as well as its specific characteristics.

4.5. Conclusion

To analyze the effects of digital technology on the transformation of the love relationship process and, ultimately, on the perception of love it engenders, we began by describing the evolution of human relationships through social networks. From the point of view of social relations, these networks provide an almost permanent interconnection between people. Furthermore, by highlighting closed social networks, they enable the formation of groups between people who communicate exclusively via the Internet, without ever having met.

This novelty in exchanges between people, now extended to the majority of humanity, is concomitant with the evolution of the couple over the last half-century in Western societies. As a result of social changes linked to material organization and the loss of collective references, couples now operate at the level of the individual rather than the family, giving rise to a new diversity in couple formats. Dating platforms have taken this evolution into account to build the architecture of exchanges within them, giving rise to an online dating market.

We set out to analyze the effects of dating sites on the love relationship dynamic, offering to distinguish in a first step the effects of digital platforms on the dimensions of the love relationship dynamic. Then, in a second step, based on a theory of love that distinguishes three parameters of the love relationship dynamic, namely intimacy; passion; commitment, we formulated the hypothesis that dating platforms could have a negative effect on commitment.

This chapter will have fulfilled its purpose, if it helps to encourage researchers to investigate further on how digital platforms modify love behavior and thus the conception of love.

4.6. References

Bankov, K. (2019). Cultural transformations of love and sex in the digital age. *Digital Age in Semiotics & Communication*, 2(1), 7–17.

Bergström, M. (2019). *Les nouvelles lois de l'amour : sexualité, couple et rencontres au temps du numérique*. La Découverte, Paris.

Bergström, M. (2021). Les usages sexuels des applications et des sites de rencontres. In *Les liens sociaux numériques*, Martin, O. and Dagiral, É. (eds). Armand Colin, Paris.

Dang Nguyen, G. and Lethiais, V. (2016). Impact des réseaux sociaux sur la sociabilité : le cas de Facebook. *Réseaux*, 1, 165–195. doi: 10.3917/res.195.0165.

Demonceaux, S. (2014). S'aimer à l'heure du numérique : la relation conjugale à l'épreuve de l'hyperconnectivité. *Sociologie et sociétés*, 46(1), 125–143. doi: 10.7202/1024681ar.

Duteille, C. (2007). L'événement de la rencontre comme expérience de rupture temporelle. *Journal des lettres & sciences humaines*, 6, 81–88.

Feybesse, C. (2015). The adventures of love in the social sciences: Social representations, psychometric evaluations and cognitive influences of passionate love. PhD Thesis, Université Sorbonne Paris Cité, Paris.

Finkel, E.J., Eastwick, P.W., Karney, B.R., Reis, H.T., Sprecher, S. (2012). Online dating: A critical analysis from the perspective of psychological science. *Psychological Science in the Public Interest*, 13(1), 3–66. doi: 10.1177/1529100612436522.

Goffman, E. (2002). *L'arrangement des sexes*. La Dispute, Paris.

Guéret, C. (2020). *Aimer c'est prendre le risque de la surprise*. Albin Michel, Paris.

Hertlein, K.M. and Webster, M. (2008). Technology, relationships, and problems: A research synthesis. *Journal of Marital and Family Therapy*, 34(4), 445–460.

Illouz, E. (2007). *Cold Intimacies: The Making of Emotional Capitalism*. Polity, Cambridge.

Le Douarin, L. (2002). Le micro-ordinateur dans l'espace conjugal. *Cahiers internationaux de sociologie*, 1, 169–201. doi: 10.3917/cis.112.0169.

Marquet, J. (2009). L'amour romantique à l'épreuve d'Internet. *Dialogue*, 186(4), 11–23. doi: 10.3917/dia.186.0011.

Miljkovitch, R. and Cohin, E. (2007). L'attachement dans la relation de couple : une continuité de l'enfance ? *Dialogue*, 175(1), 87–96. doi: 10.3917/dia.175.0087.

Minina, A., Masè, S., Smith, J. (2022). Commodifying love: value conflict in online dating. *Journal of Marketing Management*, 38(1/2), 98–126. doi: 10.1080/0267257X.2022.2033815.

Neyrand, G. (2004). Monoparentalisation et précarisation, des processus interactifs. *Dialogue*, 1, 3–15. doi: 10.3917/dia.163.0003.

Neyrand, G. (2015). L'impact conjugal du virtuel. Éclatement des façons de faire couple à l'heure d'Internet. *Dialogue*, 210(4), 59–70. doi: 10.3917/dia.210.0059.

Pagotto, C. (2016). L'amour à l'ère du numérique by Camille Pagotto [Online]. Available at: https://issuu.com/camillepagotto/docs/en_ligne.

Richter, L. (2014). Internet infidelity – A phenomenological study of critical issues, treatment & assessment. PhD Thesis, Nova Southeastern University, Fort Lauderdale.

Rosenfeld, M.J. (2017). Marriage, choice, and couplehood in the age of the Internet. *Sociological Science*, 4, 490–510. doi: 10.15195/v4.a20.

Salmela, T., Colley, A., Häkkilä, J. (2019). Together in bed? Couples' mobile technology use in bed. In *Conference on Human Factors in Computing Systems – Proceedings*, 1–12. doi: 10.1145/3290605.3300732.

Société psychanalytique de Paris (n.d.). L'impossible objet de la transmission dans l'anomie de la déliaison sociale [Online]. Available at: https://www.spp.asso.fr/textes/textes-et-conferences/psychanalyse-sciences-humaines/l-impossible-objet-de-la-transmission-dans-lanomie-de-la-deliaison-social/ [Accessed 13 June 2024].

Statista (2018). Digital market outlook statista [Online]. Available at: https://www.statista.com/outlook/digital-markets.

Sternberg, R.J. (1986). A triangular theory of love. *Psychological Review*, 93(2), 119–135. doi: 10.1037/0033-295X.93.2.119.

Tronca, L. and Sità, C. (2019). Réseaux sociaux et travail social. Un défi pour l'interactionnisme structural. *Revue européenne des sciences sociales, European Journal of Social Sciences*, 57(2), 193–227. doi: 10.4000/ress.5876.

5

AI and Smart Cities, What is the Ethical Positioning?

5.1. Introduction

The smart city concept is a response to the deep changes taking place in society, particularly the acceleration of global urbanization. This approach to the city is based on interconnecting networks by cross-referencing the data emitted by each of them. It is the advent of smart and data that makes this new urban ecosystem possible.

In addition, we need to take into account the fact that the organizational, technological and societal evolutions of today's cities are driven by their desire to be dynamic actors in terms of responding to the problems of ecological transition, as well as in terms of social issues. The smart city thus seeks to reconcile the pillars of sustainable development through a global approach, combining participatory governance and enlightened management of natural resources, with a view to satisfying the needs and expectations of institutions, businesses, associations and citizens (Beretta 2015).

In this context, one of the foundations of the smart city must be to enable its inhabitants to live in harmony with economic and social actors, by producing without depleting resources, but by regenerating them.

The aim of this chapter is to identify the main challenges and avenues of action for moving from the highly technocentric models of smart cities to an approach based on the symbiotic economy, enabling these urban territories

---

Chapter written by Franck DEBOS.

to have a positive environmental and social impact. Given that artificial intelligence is taking a pivotal role in smart city development projects, we need to start by considering the necessity to establish an ethical framework for AI.

To do this, our approach will mainly consist of:

– 10 semi-structured interviews with leaders of associations involved in environmental, social and cultural issues;

– 10 semi-structured interviews with representatives of local authorities involved in "smart" initiatives.

These interviews will be complemented by documentary research on the issues presented. The aim here is to not be overly "naive" about smart cities, but instead to take into account the risks of drift inherent in them, of which AI is a significant component. We must not lose sight of the fact that the human dimension must be at the heart of smart city design. In fact, the digitization of the city begins at the level of the citizen, their home, their neighborhood: this is where the link begins, the interaction, the knowledge of the uses of each actor and each component of the daily life of the "citizen" employee. The scope within which they live with their family, expresses themself, consume, exchange, contribute and vote.

We are, of course, fully aware of the explorative nature of our approach. Nevertheless, we hope to be able to demonstrate, through their interrelations and respective expectations, the importance of the role played by associations in the successful development of a smart city, while respecting the freedom of the citizens who make it up, which implies an ethical framework for the AI processes used for its emergence and growth.

5.2. The necessity and challenges for ethical AI

AI is one of the major issues of the age of convergent technologies, with far-reaching implications in all fields. Like previous technological revolutions, AI has the potential to transform the future for the better, thanks to new tools. However, these same tools can also bring many potential risks and challenges.

The uses of AI are becoming increasingly prevalent in our daily lives, as well as in business and society. As a result, the deployment of these technologies requires in-depth ethical reflection (Seraiocco 2020). Among the key issues surrounding AI, we can identify its incidences in terms of democratic practices (Cambridge Analytica affair in 2016), the large-scale collection of personal data, some of it sensitive, as part of targeted marketing strategies to better comprehend and anticipate behavior.

In this context, certain uses of AI have been subject to criticism through the creation of algorithmic biases (Cardon 2015) that can reinforce certain forms of discrimination, exclusion and social inequalities (Noble 2018). On the other hand, these few reflections should not overshadow the positive implications of AI on a societal level, within sectors such as healthcare, urban planning or even housing.

As a result, we see today that several institutional, political and entrepreneurial actors, as well as citizens, are insisting on the need to develop a reflection on an "ethics of/for AI" (Menissier 2019). It is hence necessary to question the preposition "of", which, depending on our interpretation, will engage distinct levels of reflection.

First of all, should we integrate the notion of ethics applied to AI into general ethics, or should we approach it in a specific way? If we consider AI ethics to be a sub-category of general ethics, then we are in the familiar framework of the emergence of a technological innovation, which implies:

– a certain level of uncertainty, due in particular to a partial degree of knowledge and mastery of this innovation;

– concern for the impact of the innovation on society and people's daily lives.

The history of the emergence and spread of innovations over the past centuries tends to show that developments in AI and underlying or associated technologies could call into question the balance between an individual's freedom of thought and enterprise, and their share of responsibility to society. Indeed, it would seem that the spread of these technologies reduces both the degree of freedom and responsibility of individuals. In this context, the ethical approach enables citizens to identify their degree of responsibility, as well as to "guide" their freedom.

Indeed, the deployment of AI, like that of digital technologies in general (also in other fields of innovation such as robotics, biotechnologies, healthcare, etc.), contextualizes ethics within a computational logic that stems from consequentialist reasoning; in other words, a process that will determine the positive or negative character of an innovation and its implementation, according to its observable consequences on the various dimensions of its environment (macro and micro), as well as on the populations targeted by this innovation. This impact can vary according to the environments and populations considered.

In this context, it is essential to adopt a plural vision of ethics, rather than the one focused solely on the consequentialist dimension. Indeed, when we integrate certain disciplinary fields such as philosophy, more specifically moral philosophy, we can identify three other categories of ethical approaches, namely:

– *deontologism*, focused on the notion of duty or internalized moral experience;

– *aretaism*, which focuses on the notion of virtue and takes into account the differences in the perception of this concept according to a given socio-cultural context;

– *axiologism*, which emphasizes cardinal values such as religious ethics.

As we can see, consequentialist logic based on utilitarianism is just one of four categories of ethics. These various forms are present in the philosophical literature, as well as in other human and social sciences, and are as legitimate and coherent as consequentialism. They are fully relevant and in fact enable us to better comprehend the ethical options associated with the development of AI. At present, the approach to AI ethics must take into account not just one category, but all four of the ethical dimensions presented above.

These few reflections are essential when designing a smart city, which will in fact make large-scale use of digital tools integrating AI during its development.

5.3. A truly smart city?

We can observe that the image of megacities in terms of quality of life for their citizens, as well as in terms of governance, is rather negative. Indeed, at least since Fritz Lang's Metropolis in 1927, in the imagery and scenarios of popular culture markers that have a major impact on our collective subconscious (films, series, novels, comics, manga, video games), megacities have been likened to a futuristic dystopia[1].

Compared to this highly degraded vision, the smart city can symbolize a restorative utopia, a kind of megalopolis resilience with enlightened governance ensuring a certain quality of life for its citizens. This may explain why, on the face of it, the smart city concept has been enthusiastically embraced for the past 20 years or so, as a symbol of the "happy" side of metropolization, a certain urban renaissance.

Historically speaking, the smart city concept follows in the footsteps of a certain technocratic urban planning style born in the 19th century and associated with the arrival of electricity. In the following century, the approach to cities was primarily based on a logic of flows and networks at different levels (water, energy, transport, human, digital), as illustrated by the Burnham plan for Chicago between 1900 and 1910. The organization of cities is also impacted by the cybernetic trend initiated by Norbert Weiner. This technocentric vision and dimension is aptly addressed by Antoine Picon, in his book *Smart cities, théorie et critique d'un idéal autoréalisateur* ("Smart cities: theory and critique of a self-fulfilling ideal"). In the same vein, we can cite two cases of attempted smart cities before their time. The first was the city of Los Angeles, which developed a similar system in the early 1960s, and the second was the "Cybersyn" project created by Anthony Stafford Beer in Chile during the presidency of Salvador Allende in the following decade (a project aborted by the 1973 *coup d'état*).

At the end of the 20th century and in the first quarter of the 21st century, the concept of the smart city emerged, with the underlying idea of interconnecting different networks by cross-referencing the data emitted by each of them. Such an urban ecosystem can be envisioned through the strong growth of smart and data, data that is retrieved through various media (sensors, meters, NFC chips, GPS, video surveillance, etc.), and which is then processed, cross-referenced and analyzed (datamining). Finally, all this data is made visible in a way that is clear and comprehensible to all citizens (at least on a theoretical level). A review of academic research and studies tends to show that these tend to focus on the technological management of smart cities (Lee et al. 2013), the impact of innovation in cities (variously

1 Examples include *Blade Runner, Judge Dredd, A Crystal Age, Elysium, Ready Player One* and *What Happened to Monday* for films, and *Cyberpunk 2077, Fallout New Vegas, BioShock* and *Dishonoured* for video games.

defined) on their performance (Kourtit and Nijkamp 2012; Komninos et al. 2013) or the analysis of urban governance issues associated with the multiplicity of actors.

This new urban approach highlights the need to interconnect networks by cross-referencing the data emitted by each of them. The aim is to make these networks "intelligent", with a view to anticipating everyday situations. It is the advent of smart and data that enables us to perceive and imagine this new urban ecosystem (Gassmann et al. 2019).

Finally, the notion of smart city refers not only to the emergence of a set of innovative urban services through the propensity of information and artificial intelligence to the Internet of Things (IOT), but also to the connectivity of all networks (telecoms, electricity, gas, water, transport, etc.). The fundamental mission is to ensure the coherence of urban growth, energy and environmental constraints and limits, with the central aim of enhancing the well-being of the population.

However, we can consider that digitization is only one component of this urban vision, and that the smart city must be approached through the set of criteria that enable a city to be economically competitive, efficiently managed and offer an attractive living environment. In this context, and taking into account work such as that presented by Giffenger et al. (2007), a city will be recognized as "smart" if it integrates the following dimensions:

– *Smart economy*: innovation is facilitated through dedicated structures and services.

– Smart people: training is funded.

– Smart governance: the city is efficiently managed.

– *Smart living*: a real quality of life is provided.

– Smart environment: the city has good environmental performance.

– *Smart mobility*: the whole package is on offer, within the framework of sustainable mobility.

These six dimensions are presented in this context as the required aspects that a city must integrate and respect in a logic of "smartification" in order to achieve the smart city status. Of course, the risk is that this position will lead to a reductionist vision of the smart city, ignoring the need for a specific analysis of each city's problems. In this context, a "smart" city must fully integrate the normative vision that is taken up by the prevailing discourse (notably political), i.e. mainly a city that is economically efficient, environmentally and energetically responsible, promoting participatory democracy and concerned with offering its citizens a quality living environment (Hollands 2008; Caragliu et al. 2009). It is important to bear in mind, however, that considering smart cities solely in terms of these criteria primarily corresponds to the need to achieve both political and territorial marketing objectives, for better positioning in rankings that incorporate these criteria.

As cities are the source of 80% of the world's GDP, we can therefore consider that the smart city ultimately questions a choice of civilization through:

– a choice to create value, prioritizing "progress" objectives associated with quantitative and qualitative indicators of success;

– a choice of quality of life and lifestyle for its current and future citizens, as well as for the associations, institutions and businesses operating in its territory;

– a choice of administrative governance in terms of territorial and state public services, as well as financial, political, economic, cultural, social and environmental (Pisani 2015; Calzada and Satyam 2017).

The concept of the smart city, originally carried by the dominant actors in the IT sector and now being joined or replaced by digital giants, explains why the latter are currently assuming the role of a global prime contractor, as illustrated by the Quayside smart district in Toronto, piloted by a subsidiary of the Alphabet group and featuring AI everywhere. In this technology-driven vision, we can see a power shift in which American electronics and digital multinationals are in the lead, but are being overtaken by their Chinese counterparts.

The underlying risk is that the political decision-makers associated with these companies will design ex nihilo an ideal city in a laboratory by experts, scripted and visualized on virtual 3D models, with little or no regard for the existing population, historical heritage or local culture, intended for elite groups. This vision gives rise to and encourages the from-scratch trend, i.e. the creation of new urban spaces from scratch, in order to exclude all the material, cultural and social complications associated with the evolution of an old city. This deliberate trend towards rapid, top-down planning is very much in evidence in emerging countries, particularly in Asia and Africa.

By way of illustration, the "Vision 2030" plan initiated by Saudi Arabia presents several smart city projects such as "The Line", which illustrates this desire to create from scratch, as well as other projects integrating cultural ambitions and tourism objectives. The kingdom plans to enhance an exceptional archaeological site in the Al Ula region[2]. In this original and monumental project for the ex nihilo creation of a city in the desert, the stakeholders insist on the cultural dimension (we find the Louvre Museum already present in Abu Dhabi, the French archaeological services, the Paris Opera, the La Fémis film school, etc.). Another noteworthy project is Jean Nouvel's design for a luxury hotel to be integrated into the rock, following the example of the Nabataean buildings.

In France, the Nice Côte d'Azur metropolis was singled out in 2014 at the eco-citizen energy awards in Paris, thanks to projects focused on smart grids (*Ecofamilies*, for which the author of this chapter was the project leader for the research and of City-Opt)[3].

5.4. Directions and aspirations for smart cities

Analysis and synthesis of semi-structured interviews with association representatives and local authority decision-makers have identified three possible scenarios.

2 The site, a UNESCO World Heritage Site, includes 111 monumental tombs, vestiges of the Nabataean civilization (4th century BCE), as well as the more modest Petra site, which attracts many tourists. This is a colossal project, with a budget of between 50 and 100 billion dollars and the aim of attracting 2 million visitors a year. In 2018, France was entrusted with the archaeological excavation of Madain Saleh (where French researchers have already been present for 15 years) and the development of the cultural park, corresponding to a budget of 40–80 billion euros. This is the background to the creation of the French Agency for the Development of the Al-Ula Archaeological Site, headed by Gérard Mestrallet, former Chairman of Engie.

3 "La métropole primée pour sa stratégie énergétique", *L'Avenir Côte d'Azur*, no. 2191 of August 22, 2014.

5.4.1. *A digital business city: profit center*

This vision of the smart city is predominant among institutional decision-makers. Indeed, the goal of collective enrichment of the city or region, as a factor of development and investment capacity, is the one most emphasized by local authorities wishing to achieve international smart city status. Here, the priority is clearly economic, and the environmental and social dimensions are secondary, despite announcements to that effect. This context implies going beyond conventional infrastructures. The aim is to achieve a high-tech, globalized edge, illustrated by steel and glass towers and a building stock that will lead to gentrification, symbols of a certain avant-gardism, with the consequence that the original population will be financially and culturally excluded from this urban modernity, in favor of an external population with a rather high standard of living, for expatriates, what David Goodhart calls people from anywhere (Fourquet 2021).

This vision of the smart city, associated with a goal of totally technodigital governance, and sought by the vast majority of institutional decision-makers surveyed, currently predominates (Guarrigues and Blanc 2016). In this highly complex digital city model, the aim is to eliminate all failures and hazards (particularly those of human origin), to digitally integrate all the functions and processes of city governance, and to automate analyses and decisions in real time to make them more agile. This "techno-solutionist" vision, in which AI plays a decisive role, has been developing over the past 15 years, and is proceeding in two phases: a global coverage of sensors in a logic of panoptic surveillance of both technical and human activities, followed by the provision of software platforms to automatically process the billions of data collected. In this context, decisions are top-down and prioritize the system's functional rules, rather than the concrete needs and expectations of its inhabitants, who are considered passive beings whose behavior must be standardized and modeled to ensure that nothing impedes the smooth running of this urban entity. Here, smart governance implies transferring the diagnosis of the city's problems and the resulting decisions as quickly as possible to algorithms that are supposed to be capable of resolving everything almost instantaneously. Smart city 3.0 is already on the horizon, thanks to increasingly "profound" (or even conscious in the near future?) artificial intelligence, which merges and processes all Big Data in order to optimize urban management solutions in real time, leaving little room for human decisions, those of institutional actors and the free expression of citizens. Even if local authority representatives are currently making their own decisions (respondents to our survey with this profile all said so), it seems highly likely that AI designers will develop tools with such a high level of responsiveness that individuals will no longer be able to compete, let alone control.

The greatest risk, however, is not technological, but rather socio-cultural, in the form of a globalized, "techno-dehumanized" model. Taking into account the criteria of such a smart city implies the eventual abandonment of our own culture and history. We are faced with a standardized, impersonal model where all cities look alike, and where the only distinctive features will be ever-taller mineral towers. Smart cities are cold witnesses to technological prowess, with no regard for environmental and social issues. Instead, they are representative of ostentatious wealth, reminiscent of the imagery of a blissful future often preceding a catastrophe, as seen in popular culture narratives (films, series, video games, comics, etc.). Indeed, in a logic of eliminating the hazards and disturbances caused by the irrationality of its inhabitants, this city will develop global surveillance, becoming a jail city intent on standardizing and teleguiding the lives of its citizens. These considerations seem to be underestimated, if not absent from the responses of the institutional decision-makers interviewed.

More or less consciously, this smart city model, emphasized by companies dominating the digital market, as well as by start-ups, requires very substantial high-tech investments aimed at an elite of digital nomads who are globalists, innovative, ultra-connected and fond of objects incorporating an ever-growing IOT. In this context, and as we can already observe in several metropolises, the high cost of living is increasing and excluding the majority of citizens with more modest incomes, whether working, students or retired, creating an unequal gentrification movement.

For the past 20 years or so, global observations and analyses in this area have pointed to a process of city modernization that is vertical, voluntarist and authoritative. In this context, the goals, priorities, technological choices and implementation processes applied are the work of, and under the control of, the cities' political leaders and their staff of specialists, consultants and managers. In this model, stakeholders are absent from any kind of decision-making, whether they be citizens, associations, businesses (especially VSEs/SMEs) apart from those participating in the development of the smart city, intermediary bodies, etc., in fact all those "who are and do" the city. Everything is controlled by political decision-makers, their experts and advisors, who see themselves as the only ones capable of making the best decisions for managing their technocity. More than the nature and type of modernization, it is the transformational approach and implementation of technological innovations that are called into question.

In the end, our approach is the opposite of that of marketing (which is paradoxical when you consider the profile of the companies involved in creating a smart city). Indeed, decision-makers make little (if any) attempt to reach a consensus between the aspirations of the city's stakeholders and the quasi-imposed offer of political decision-makers, who are driven by the objectives of image, national and even international prestige, and the creation of a digital-centric economic dynamic.

5.4.2. *Ecotechnocity or ecolife city*

This other vision of the smart city is also common among institutional decision-makers, as well as shared by a number of associative actors focused on sustainable development and ecological transition. We can, of course, welcome the fact that the environmental dimension, particularly through the ecological transition, is seen as an essential element of urban development. It is nevertheless important to question the type of objective that underpins this sustainable development approach, the choice of criteria, the approach, the associated indicators, the preferred targets, etc.

Do we want a carbon-neutral city with a neutral or positive environmental impact, but at the cost of permanent control of its citizens in a logic of restrictive, even punitive ecology? Or do we want an "organic" city, a "bio-city" with a similar goal, but one that respects the objectives of sustainable development? We could, for example, take into account the UN's SDGs to design an ecosystem in which nature and quality of life have a prominent position. This was a point of debate among the social actors who took part in our study. In this context, and accentuated by an ambient ecoanxiogenic discourse, citizens will directly perceive the restrictive elements associated with this eco-environmental goal (reduction of consumption, travel by individual vehicles, selective sorting, sobriety logics, etc.) and less their well-being, their quality of life, which may be negatively impacted by these measures.

It is also important to point out that, with a view to boosting their profile and international influence (and this is reflected in some of the responses from those questioned), many local authority managers are talking about the concept of a sustainable city. The latter is packed with technological innovations and AI to assess and anticipate risks, regulate mobility flows, optimize energy expenditure and reduce polluting emissions and carbon impact. However, in this urban model, the temptation is great to try to model and control the behavior of its inhabitants, visitors, businesses, associations, etc. The danger is therefore to drift towards a restrictive, negative conception of ecology, whose central mission in this case would be to reduce the environmental impact of its citizens through high-tech solutions in particular, whereas we need to think of the city in a regenerative logic, putting its inhabitants at the center of reflection in the service of a "revitalizing" ecology (Ingallina 2019). In this context, we also need to facilitate the return of animals, considering them as true inhabitants of this regenerative ecosystem in the same way as humans.

5.4.3. *Ethic and the human city*

The vast majority of respondents, who belong to associations with a social, environmental or cultural mission, have a more human-centric vision of the smart city. The central element here is the quality of life experienced and felt by its inhabitants, which is more difficult to measure, as it is more subjective and heterogeneous by definition. The objectives of economic development and performance, as well as those associated with the environmental impact of the smart city, are also present, but it must above all be pleasant to live in and sociologically "lovable" (Moreno 2013). However, even if this approach is widely shared and diffused among associative actors (and surely the majority of citizens), it receives little media coverage and very little presence in dedicated events, because the high-tech, predominantly digital dimension, with complex AI, is not a priority. We find this approach in Northern European cities such as Helsinki, as well as in cities in the "global south" such as Medellin[4], which was awarded the

4 Medellin, after a chaotic period associated with the drug cartels (notably Pablo Escobar), has succeeded in reinventing itself by refocusing on genuinely meeting the expectations of its population. In fact, almost a quarter of the costs involved in modernizing the city are directed towards improving the quality of life in poor neighborhoods which, until now, had been completely abandoned. Most of these actions are focused on improving the daily lives of residents, and do not automatically incorporate technological innovations. Primacy is given to exchanges with the population, and to strengthening its involvement in urban development projects in order to restore the quality of life and social peace. This is a far cry from the logic of a city over-equipped with sensors that analyze the behavior of the population in real time, with a view to security and pseudo-prediction. In terms of

Lee Kuan Yew prize at the 2016 World Cities Summit. In this model, progress and performance are expressed not through ultra-connected glass towers, but in a higher quality of life, greater solidarity and a significant drop in urban violence. In this context, the issues to be resolved focus on the day-to-day lives of the city's residents, its employees, and the companies and associations that form part of it, and technological solutions must respond to these human challenges with the most personalized answers possible. Every resident has the right and opportunity to express their opinion, and to make proposals concerning all development projects in the city.

In addition to budgetary constraints, the smart solutions developed integrate sociological issues, as these innovations must be perceived as factors of progress in terms of lifestyle for both citizens and the city agents who will have to implement them. As indicated by all the social actors who took part in the interviews, co-creation and co-modernization practices are commonplace here, with bottom-up logic, social progress, collaborative participation, solidarity and fair treatment as key words. Here, the smart city is first and foremost a human community, rather than a place of submissive implementation and technological experimentation. Here, technological innovation and, more specifically, AI, are seen as mere supports for overcoming constraints and achieving the objectives of social innovation. This is why, in this context, mid-tech (which has already proved its worth and is easily transposable) or even low-tech will often be preferred to high-tech smart solutions saturated with AI. It will therefore be more difficult, but not impossible, to develop public–private partnerships, as financial investors and companies are naturally more attracted to a technocentric smart city project incorporating numerous high-tech innovations.

As indicated by the social actors interviewed and some institutional decision-makers, the primary aim of this alternative vision of the smart city is to improve collective rather than individual quality of life, and to preserve the sociological equilibrium of the entire population concerned. According to the social actors interviewed, this new urban ecosystem needs to be more equitable, to develop the inclusion of all the social groups that make it up, particularly those at risk of marginalization or even exclusion. The dynamic driving force behind this category of smart city can be summed up as the combination of an ethical dimension and a certain socio-economic pragmatism, in a logic of fair treatment, social peace and enhanced quality of

results, Medellin has recorded a 90% drop in serious crime, and a reduction in extreme poverty from 8% to 3%.

life. For all respondents from the associative fabric, the smart city must be "sustainable", and therefore respond to ecological issues, but at the same time be capable of offering a pleasant, comfortable, dynamic and lively living environment, thus creating a strong appeal, a desire to stay and to participate in the influence of this city.

5.5. Conclusion

A city will be able, if not to optimize, then to increase the performance of its flows through the extensive use of technology, particularly associated with digital technology, under the aegis of centralized urban governance, in partnership with major corporations. In smart cities, this is currently the dominant trend, with little or no consideration given to socio-cultural issues, in favor of high-tech devices containing elaborate AI processes. What is more, this situation is reinforced insofar as the media, congresses and trade fairs give pride of place to this technocentric vision emphasized by the digital giants (Wolframm 2012). By way of smart city, or even smart territory, we will have, in the majority of cases, only a succession of modernization actions associated with a certain political opportunism, actions which will not necessarily be coherent and which will not create in their citizens, and as a result, a clear image of the city as well as a desire to live and/or stay there.

These increasingly digitized smart cities are multiplying the number of data sensors, individual recognition software (biometric, facial, verbal), analysis of written and visual data from social networks, as well as numerous cross-referenced files, and are thus progressively moving towards global and ubiquitous surveillance, both collective and individual. This raises a number of ethical questions, including how to manage this data and how to use AI. Indeed, some of this data will be used in the form of global, anonymous statistics. On the other hand, a significant proportion of this data falls within the "intimate" private sphere, which must be respected, but which is also very attractive to companies and political decision-makers. As a result, cities are currently emerging – and have been for at least a quarter of a century – that incorporate numerous digital functionalities, without any thought of having been given to their purpose and powers.

In terms of the three orientations associated with the interviews conducted, and also found in the literature on this subject, the choice essentially arises between a *digital business city* optimizing its various flows and data through constantly perfected high-tech technologies, integrating environmental issues, but to the detriment of its humanization, and an ethical human city focused on the expectations of its citizens, and hence on the implementation of social innovations at the organizational level, as well as in terms of relations and collaboration between its inhabitants and stakeholders, aided in this by existing or innovative technological supports that are well mastered by its inhabitants (Haentjens 2021).

We must not lose sight of the fact that the choice of positioning and orientation of the smart city will condition the evolution and attractiveness of this urban ecosystem to citizens. However, as Greenfield (2013) states, in order to prevent the city from being in the hands of *datacrats*, all citizens worldwide need to think carefully and collectively about the place they wish to assign to the various technologies (especially digital ones) associated with the smooth running of their cities, and their impact in terms of satisfying their expectations, needs and desires. It is also worth pointing out that all those with a certain degree of mastery information and communication technologies (ICTs) have a duty to raise awareness and educate all citizens in this area, with a view to facilitating decisions that will help build a smart city centered on human and social dimensions. The essence of a smart city is not the constant pursuit of modernity through the continuous implementation of high-tech innovations, but the well-being of its population and its stakeholders. Innovations, whether disruptive or continuous, must of course be present, as long as they are integrated into the social innovation process that has been put in place. The logic here is not "technomarket", but rather co-modernization (Townsend 2014).

All of our associated reflections, and those stemming from our interview approach, tend to show that the future of smart cities (and more broadly of smart territories) must be based on an interactive and co-creation logic aimed at integrating, not only citizens, but also associations and companies as actors and "shapers" of their cities, which calls for reflection on the ethical framework that the processes used and based on AI must respect (Vanin and Cyrulnik 2019).

5.6. References

Beretta, J. (2015). Les caractéristiques d'une ville intelligente. Dossier Smart Grids de la CRE (Commission de Régulation de l'énergie).

Calzada, I. and Satyam, A. (2017). *The Smart City Transformations*, 1st edition. Bloomsbury, London.

Caragliu, A., Del Bo, C., Nijkamp, P. (eds) (2009). Smart cities in Europe. In *Creating Smarter Cities Conference*. Edinburgh Napier University, Edinburgh.

Cardon, D. (2015). *À quoi rêvent les algorithmes ?* Le Seuil, Paris.

Fourquet, J. (2021). *La France sous nos yeux, économie, paysages, nouveaux modes de vie*. Le Seuil, Paris.

Gassman, O., Bohm, J., Palmie, M. (2019). *Smart Cities: Introducing Digital Innovation to Cities*. Emerald Publishing Limited, London.

Giffenger, R., Fertner, C., Kramar, H., Kalasek, R., Pichler-Milanovic, N., Meijers, E. (2007). Smart cities – Ranking of European medium-sized cities. Centre of Regional Science, Vienna [Online]. Available at: http://www.smart-cities.eu [Accessed 25 November 2010].

Goodhart, D. (2017). *The Road to Somewhere: The Populist Revolt and the Future of Politics*. C Hurst & Co Publishers Ltd, London.

Greenfield, A. (2013). *Against the Smart Cities*. Do Project, New York.

Guarrigues, A. and Blanc, S. (2016). La Smart city à la recherche de modèles économique. Club Techni.Cité [Online]. Available at: http://www.lagazettedescommunes.com/441803/la-smart-city-a-la-recherche-de-modeles-economiques/#sommaireDossierTarget.

Haentjens, J. (2021). *Smart city, ville intelligente : quels modèles pour demain ?* La Documentation française, Paris.

Hollands, R. (2008). Will the real smart city please stand up? *City*, 12(3), 303–320.

Ingallina, P. (2019). *Ecocity, Knowledge city, Smart city : vers une ville écosoutenable ?*, 1st edition. Septentrion, Paris.

Kominos, N., Pallot, M., Schaffers, H. (2013). Special issue on smart cities and the Future Internet in Europe. *Journal of the Knowledge Economy*, 4(2), 431–446.

Kourtit, K. and Nijkamp, P. (2012). Smart cities in the innovation age, innovation. *The European Journal of Social Science Research*, 25(2), 93–95.

Leea, J-H., Phaalb, R., Lee, S-H. (2013). An integrated service-device-technology roadmap for smart city development. *Technological Forecasting and Social Change*, 80(2), 286–306.

Menissier, T. (2019). *Quelle éthique pour l'IA ? Naissance et développements de l'intelligence artificielle*. Académie Delphinale, Grenoble.

Moreno, C. (2013). Les défis de la smart city humaine. In *Colloque Smart City Exhibition 2013. Communication, Quality and Development in Smart Cities*, Bologna.

Noble, S. (2018). *Algorithms of Oppression: How Search Engines Reinforce Racism*. NYU Press, New York.

Picon, A. (2015). *Smart Cities. Théorie et critique d'un idéal autoréalisateur*. B2, Paris.

Pisani, F. (2015). *Voyages dans les villes intelligentes : entre datapolis et participolis*. Netexplo Observatory, Paris.

Seraiocco, N. (2020). *Résistances et contrepoids face à l'IA*. Relation, Montreal.

Simondon, G. (2012). *Du mode d'existence des objets techniques*. Aubier, Paris.

Townsend, A. (2014). *Smart Cities: Big Data, Civic Hackers, and the Quest for a New Utopia*, 1st edition. W.W. Norton & Company, London.

Vanin, L. and Cyrulnik, B. (2019). *Smart City, une autre lecture de la ville*. Ovadia, Paris.

Wiener, N. (2014). *La cybernétique : information et régulation dans le vivant et la machine* (posthumously translated by Le Roux, R., Vallee, R., Vallee-Levy, N.). Le Seuil, Paris.

Wolframm, M. (2012). Deconstructing smart cities: An intertextual reading of concepts and practices for integrated urban and ICT development. In *P. REALCORP Proceedings*. CEIT Alanova, Vienna, 171–181.

6

The Smart City and AI: Opportunities, Challenges and Expectations for Businesses

6.1. Introduction

The smart city concept is intended to respond to the profound changes taking place in society, particularly the acceleration of global urbanization. From 54% in 2014, it is expected to reach 70% by 2050, i.e. 20% more than at present (Schneider Electric 2014). For the dominant telecoms and digital actors, smart city development projects represent a market valued at hundreds of billions of dollars (Guarrigues and Blanc 2016).

This notion remains unclear (Wolfram 2012). It is the subject of several approaches by network operators, technology providers and, from a political point of view, by the self-promotion of certain cities (Hollands 2008).

Academic research and studies tend to focus on the technological management of smart cities (Leea et al. 2013), the impact of innovation in cities (defined in a wide variety of ways) on their performance (Kourtit and Nijkamp 2012; Komninos et al. 2013) or the analysis of urban governance problems induced by the multiplicity of actors. There is still little work on the impact of businesses on smart cities.

The idea behind this new approach to the city is to interconnect networks by cross-referencing the data emitted by each of them. The aim is to make

---

Chapter written by Franck DEBOS and Maria Daniella LANCINI.

these networks "intelligent" so that they can predict everyday situations. It is the advent of *smart* and *data* that make it possible to envisage this new urban ecosystem (Gassmann et al. 2019).

In addition, while the smart city essentially exploits the data collected by physical sensors (in order to know and monitor the state of networks), information from social networks is often under-exploited and could constitute another promising source for understanding the city. Already, local authorities almost systematically use their Facebook, X, Instagram or Snapchat accounts to quickly relay any useful information to connected residents (traffic, roadworks, events, exceptional situations). As for local residents, they have also taken to social networks to create communities in their towns, for example, in sports (*Strava, Runtastic*) or in associative circles (event organization).

We also need to take into account the fact that the organizational, technological and societal evolutions of today's cities are driven by their desire to be dynamic actors in terms of responses to the challenges facing sustainable development and, in particular, climate change. The smart city thus seeks to reconcile the pillars of sustainable development through a global approach combining participatory governance and enlightened management of natural resources, with a view to satisfying the needs and expectations of institutions, businesses and citizens.

Against this backdrop, many companies have taken part in this evolution of the city, either through their natural positioning (Orange, Lafarge, JC Decaux, start-ups, etc.) or through their technical capacity to drive this change (IBM, Cisco, Thales, etc.). However, not all companies have taken the full measure of this phenomenon.

And yet, it seems obvious that local authorities (especially cities) and businesses are indeed linked to improve everyday life and meet the needs of the "citizen/user/consumer'-actor". We are talking about the same person, whose uses, expectations, demands and even contradictions have evolved considerably since the development of the Internet and cell phones. And yet, whether in the many conferences on the subject, or in the articles describing in detail how to make a smart city a success, there is little mention of the central position that companies must hold, even though their role is essential. By way of illustration, it is hard to imagine that telecom operators, Google Cars or IBM algorithms alone will resolve traffic congestion in urban metropolises, if businesses do not adapt their opening hours or logistics activities at the same time.

The aim of this article is to identify the main expectations and common challenges facing French companies with regard to smart cities, whatever their sector of activity or size. These considerations will enable us to define their decisive importance in the transformation of a "traditional" city into a smart city.

To this end, our approach has mainly consisted of:

– 10 semi-structured interviews with business leaders from all sectors (interviews conducted with participants at the "Smart Cities and Territories Expo", October 20–21, 2021, Paris Expo, Porte de Versailles);

– 10 semi-structured interviews with representatives of local authorities involved in "smart" initiatives.

Independently of this, documentary research on the challenges presented will be carried out in addition to these interviews.

Nor will we take an overly "naïve" view of smart cities and the role of businesses in their development. We will also take into account the risks of these smart cities and the smart businesses that are part of them drifting off course and becoming jail cities.

We must not lose sight of the fact that the human dimension must be at the heart of the design of a smart city and that the digitalization of the city begins at the level of the citizen, their home, their neighborhood, the boundary within which they live with their family, express themselves, consume, exchange, contribute and vote. The transformation of the company will also begin at the level of the employee, their office, their department and their management.

We are, of course, fully aware of the exploratory nature of our approach, and we nevertheless hope to demonstrate, through their respective motivations, interrelations and expectations, the importance of the role played by companies in the successful development of a smart city while respecting the freedom of the citizen-consumers.

6.2. The sociotechnical context of the smart city

Before delving further into the subject, we would like to briefly outline the socio-technical context of the smart city. Since the beginning of the 21st century, the large-scale development of digital technology has had a considerable impact on the structures of organizations, including local authorities, as well as on their managerial practices and the relationships between their stakeholders, such as employees, users, companies, associations and institutions (Dorn 2017; Brasseur and Biaz 2018). This logic of increasing digitalization gives rise to antagonistic visions of the future of organizations. For some, it brings with it a wave of job destruction and will contribute to rising unemployment in a context of increasing control over individuals (Brynjolfsonn and McAfee 2011). On the contrary, for others, it creates new functions and innovative professions, resulting in a responsible economic upturn and individual fulfillment. The key challenge is to achieve a technological humanism (Simondon 2012) that prevents the individual from being overtaken by the technologies they have designed.

Chronologically speaking, we can identify several stages in the process of digitizing organizations (Cazals 2015). The first stage concerns the 1980s, with the integration and development of microcomputing and client/server infrastructures. Then, from 1990 onwards, came the "World Wide Web", which met the need to centralize information that had hitherto been little taken into account and was often used by a constantly changing workforce (Flichy 2011). At the end of this decade, the concept of the smart city emerged. The third stage corresponds to the 2000s, when organizations immersed themselves in digital technology, with the central objective of controlling data flows and transfers. In this context, organizations and individuals alike have access to software that enables them to share and centralize information. Over the years, the concept of the smart city has been gaining ground. Closer to our uses and issues at this level, the decade of 2010 is considered to be a period of almost complete digitalization of organizations, in a mobiquitous logic giving the possibility of new uses and instructing new data to be managed (El Hadi and Arsenault 2011). These years have seen strong growth and internationalization of the smart city concept, with companies playing an increasingly important role.

We can also observe that the notion of smart cities refers to the emergence of a field of innovative urban services that deals with the propensity of information and artificial intelligence (AI) to integrate numerous objects, as well as the connectivity of all networks (telecoms, electricity, gas, water, transport, etc.). The fundamental objective is to ensure the compatibility of urban development, energy and environmental constraints, and increased well-being for the population.

As mentioned above, the concept of smart cities is increasingly being used and experimented with internationally. One example is the Songdo business district in South Korea, which represents the model of the connected city, with cameras that scan the streets to adjust lighting according to the number of passers-by, traffic that is monitored by analyzing data from radio-frequency identification (RFID) chips embedded in cars and buses, energy and water consumption and so on. Closer to home, the Nice Côte d'Azur metropolis won an award in 2014 at the *Energies écocitoyennes* awards in Paris, thanks to projects focused on smart grids (*Ecofamilies*, for which the author of this chapter was the project leader for their research and their role in *City-Opt*)[1].

Among the earliest studies of smart cities are those by Mahinzan (1999), who discusses Singapore's strong economic development.

According to this author, a city's intelligence lies in the way it integrates information and communication technology (ICT) into all aspects of its economic life. The central objective is to jointly meet the challenges of economic growth and the development of social well-being, knowledge and human capital. The notion of a smart city cannot therefore be reduced to the digitization of its infrastructures and services (Haentjens 2021), combined with various categories of AI that are more or less developed according to the given context. From this perspective, the notion of smart city is strongly linked to its technological content. A city that is all the more "smart" for being digitized and "stuffed" with AI. This vision is widespread and forms the basis of many smart city rankings.

In this context, it is important to try and grasp all its aspects, especially as the approach to this concept varies from a highly technocentric vision[2] to a more "humanizing" approach.

1 "La métropole primée pour sa stratégie énergétique", *L'Avenir Côte d'Azur*, no. 2191, 22 August 2014.

2 By way of illustration, IBM has designed a city "intelligence index" that enables it to measure the performance of each city according to seven criteria: urban public services, "citizens" (education, health, safety, etc.), business, transport, communications, water and energy.

The cities of the 21st century must take into account the logic of metropolization, the growing complexity of the urban reality, the increasing role of digital technology and AI and the need to contain the drift of local public spending. Against this backdrop, the smart city becomes the place where urbanization challenges and opportunities can be met.

To meet the challenges of the future – quality services and infrastructure, value creation and residents' growing desire for participation – the smart city is developing digital technologies, integrating more and more AI into urban services to make them more efficient. It is also experimenting with new public–private collaborations to increase economic spin-offs and involving residents in the "making" of the city. The smart city thus goes beyond urban service management historically organized in "silos" to privilege a networked approach, in which the use of digital resources becomes essential.

In this context, and in accordance with the French Data Protection Act no. 78-17 of January 6, 1978 (consolidated version as of January 24, 2006, amended by Act no. 2004-801 of August 6, 2004), personal data is "anonymized". All or part of this data may be made available to the public. With regard to data, the challenge for the smart city is to prevent various events by extracting knowledge from a large quantity of data. The resulting information will be returned to citizens in the form of maps, graphs, animations, videos, articles, etc., via smartphone applications, software, etc.

Historically speaking, the smart city follows in the footsteps of technocratic urban planning in the 19th century, which coincided with the arrival of electricity. In the twentieth century, cities were increasingly conceived in terms of flows and networks at various levels (water, energy, transport, human, ICT), right from the start, as shown by the Burnham plan for Chicago between 1900 and 1910. The organization of cities also draws on the cybernetics movement (Norbert Weiner). Antoine Picon's *Smart Cities: Theory and Criticism of a Self-Fulfilling Ideal* sums up this highly technicist vision. To echo these reflections, we can cite examples of smart cities before their time, such as the city of Los Angeles, which set up such a system in the early 1960s, or the Cybersyn project created by Anthony Stafford Beer in Chile under Salvador Allende in the following decade (a project aborted by the 1973 *coup d'état*).

At the end of the 20th century, and in the 21st century, the notion of the smarter city and then the smart city emerged, with the idea of interconnecting networks by cross-referencing the data emitted by each of them. It is worth noting here the role played by former US President Bill Clinton, who in 2005 initiated the idea of a smart city with *Cisco* Chairman John Chambers. Bill Clinton foresaw the interrelationships between massive urbanization and the rapid spread of information technologies and wanted a strong actor in the American economy to position itself in this diffuse market. In return, *Cisco* developed a preliminary research process with an endowment of $25 million. It was also at this time that Anthony Townsend published a key work, *Smart Cities: Big Data, Civic Hackers, and the Quest for a New Utopia*, which laid the foundations for a wide-ranging and plural reflection on the smart city and the place of AI.

It is the advent of “smart” and “data” that makes this new urban ecosystem possible. In this context, data is collected via a variety of media (sensors, meters, near-field communication (NFC) chips, GPS, video surveillance, etc.). It is then processed, cross-referenced and analyzed (datamining). Finally, they are rendered in a way that is visible and comprehensible to all citizens, at least on a theoretical level.

However, we can consider that digitization is only one component of this urban vision and that the smart city can be understood as the totality of elements that make a city economically competitive, efficiently managed and offering an attractive living environment. From this point of view, and based on work such as that of Giffenger et al. (2007), a city will be considered “smart” if it becomes a crucible for innovation through dedicated structures and services (smart economy), injects funding into training (smart people), is efficiently managed (smart governance), offers a real quality of life (smart living) and good environmental performance (smart environment) as well as sustainable mobility (smart mobility). In this case, these six dimensions are presented as obligatory points of passage that a city must invest in as part of a “smartification” logic in order to be consecrated a smart city. The risk is that this position leads to a reductive vision of the smart city that is not based on a specific analysis of the challenges facing each city – challenges in which companies have a significant impact, and which are more relevant in terms of research at this level. In this context, a “smart” city should absolutely correspond to the normative vision developed by the prevailing discourse (particularly political), namely: a city that is economically efficient, environmentally and energetically responsible, that favors participatory democracy and that is concerned to offer its citizens a quality living environment (Hollands 2008; Caragliu et al. 2009). To consider smart cities solely in terms of these criteria is above all to satisfy both political and territorial marketing objectives, in order to be well positioned in rankings that integrate these dimensions. In this case, companies are more likely to be used as stooges in the political discourse to enhance the "smart" image of the city in question, without necessarily developing their actual interactions with the latter.

The Humanities and Social Sciences have also embraced the notion of the smart city, reflecting on and questioning similar concepts such as the connected city (Dutton 1997), the informational city (Castells 1996) and the sustainable city (Barnier and Toucoulet 1999; Jollivet 2001). All these definitions, approaches and analyses of the smart city reflect great diversity. We can cite, for example, research on the role of labels in the attractiveness of cities (Eveno 2018), the evolution of the relationship between capitalism and urbanization (Languillon-Aussel 2018) created by the transformation of production logics from the traditional city to the smart city, or the growing weight of the private sector in urban policymaking leading to a certain permeability of the boundary with the public sector, notably through big data.

However, it is possible to distinguish four main approaches to the smart city. Firstly, a technical approach to digital technology, in which AI plays a prominent role, sees the smart city as a means of rationalizing the urban environment and its management. This is followed by a sociological approach to science and technology, which sees the smart city as a way of increasing the complexity of city governance as a result of the uncontrolled effects of the spread of digital technology, AI and its uses. Thirdly, there is a more citizen-centric approach, which tends to see the smart city as a place for the renewal of democracy.

With this in mind, planners and researchers are paying particular attention to civic tech, whose stated aim is to strengthen the democratic link between residents and public authorities. More broadly, they are also interested in the participatory mechanisms made possible by digital tools, which implies considerable reflection on the ethics of AI. Finally, a critical approach that places the smart city within the dynamics of the global economy and suggests that the resulting projects contribute to the perpetuation of a global frame of reference for economic growth or, in other words, to the stability of capitalist logics.

6.3. Smart cities and sustainable development

The smart city also seeks to reconcile the pillars of sustainable development through a comprehensive approach combining participatory governance and enlightened management of natural resources, with a view to satisfying the needs and expectations of institutions, businesses and citizens (Beretta 2015).

This urban concept underpins the response to a number of challenges, six in number according to Moreno (2013). These are presented in Table 6.1, along with the limitations inherent in each of them.

Challenges	Limitations
Making networks more efficient and functional (AI, smart grid, sustainable development).	We need to take into account the limits associated with this issue, which is the guarantee of data anonymity, with the risk of being nothing more than a citizen-object that we trace. There is also the problem of access to public data (open data logic), which is a good thing, provided we have the necessary tools to process it.
Improving the daily lives of citizens and businesses, producing public services via new digital interfaces integrating AI.	This challenge must be balanced by the fact that in France, not all citizens are equal when it comes to technology. Indeed, only 73% of French people have a smartphone, what of the remaining 27% (Coëffé 2017).
Create digital citizen participation systems.	The question arising from this issue is that it is always the same people who participate and benefit. Can technology, particularly when combined with AI, really strengthen ties, reinforce a sense of belonging and actively involve all citizens?
Enhance environmental quality through intensive use of smart grids.	There is also the question of the environmental cost of the technological devices associated with this issue, whereas the implementation of environmental awareness and education campaigns could be less costly and more effective.
Making e-administration more efficient is one of the other key challenges of the smart city (e-administration, one-stop shopping, universal accessibility).	By way of illustration, according to the plan presenting the UK's online public services strategy, the same service developed via digital technology costs 50 times less than face-to-face service, 30 times less than by postal exchange and 20 times less than by telephone exchange (Guillau 2015). In the

Challenges	Limitations
	medium term, however, we need to determine whether this inflation of digitized administrative services will not be to the detriment of the user or SMEs and VSEs.
Respond to security issues using predictive AI.	One example is New York City's Smart Data project with *WindyGrid*. This tool provides a unified view of operational information through a single, user-friendly graphical interface, enabling users (citizens, security guards, etc.) to perform queries and receive automatic alerts and updates. In the same vein, *PredPol* software anticipates the location of crimes by analyzing statistics, enabling the police to reinforce their manpower in these high-risk areas, not without an ulterior budgetary motive. In this security context, the risk is to create tools that could be hostile towards populations deemed "undesirable" even though they use the city. Why not imagine, for example, a bench capable of recognizing the homeless and repelling them by bending down? There is also the risk of ever-increasing surveillance against the will of citizens.

Table 6.1. *Moreno's six key challenges*

In the not-too-distant future, we can well imagine city leaders with an extensive network of sensors capable of tracking everything from rainfall and traffic jams to the movements of individual citizens. They acquire an unprecedented level of omnipotence, through remote control of infrastructures and the instantaneous dispatch of transponders. Order is thus maintained in this dystopian vision of an overtly paternalistic future, as narrated by Anthony Townsend (2014). This dark vision of the development of smart cities is based on technical supports and pathways that are already at work today. A smart city is naturally endowed with a strong surveillance potential insofar as it aims to draw on masses of data, including personal data, to improve urban life. In this context, AI takes on major importance without any ethical framework.

This thinking is directly in line with the vision of the smart city as conceived by major operators such as Cisco, IBM, EDF, Engie (formerly

GDF Suez) and Veolia. They see the city as an ecosystem that is managed via centralized infrastructures whose algorithms, incorporating AI, analyze millions of pieces of data from networks, such as smart grids, and also via sensors connected to everyday objects, all linked to the Internet (the Internet of Things). The question of surveillance arises independently of data security issues. Nor does it obviate the fact that everyday technologies, starting with smartphones, already enable massive surveillance of their owners, citizens or employees of an organization. This surveillance logic will extend to the very design of cities. In France, experiments in this direction have already begun. For example, smart grids have been installed in certain districts of the cities of Issy-les-Moulineaux and Le Havre.

Everything hinges on the value of the data provided by citizens and recovered by the other two major actors, i.e. companies and public authorities. In these times of budget cuts, public–private partnerships seem inevitable for major investments.

We could also mention the dynamic congestion charging systems in London and Amsterdam, based on extremely intrusive surveillance technologies or the compulsory geolocation system in the Netherlands. In the case of the latter, the complexity and intelligence of a system do not exempt us from questioning its values, and the Dutch forced their legislator to reconsider this choice.

The American National Security Agency (NSA) eavesdropping scandal and questions about the "eavesdropping" aspect of the French military programming law also raise questions about potential abuses by public authorities. We can well imagine that, in the near future, the state will demand travel data from smart city operators.

At the same time, France's CNIL (National Commission for Information Technology and Freedom) has issued an opinion on the use of RFID chips. In the meantime, the technologies continue to be deployed, as the low fines that this body can impose for non-compliance (€150,000) are a drop in the ocean compared with the billions at stake.

We are, in fact, addressing the limits of technosolutionism, i.e. the belief that technical devices alone are the answer to city problems, leaving aside other parameters and, in particular, the human factor. This approach underpins many smart city initiatives, which ultimately ignore logic and simple common sense. For example, do we need sensors and AI to enhance the value of soft mobility such as cycling, or, on another level, e-administration and teleworking? Couldn't the solution lie in educating, exchanging and raising citizens' awareness of these issues, particularly in the age of social networks and Web 2 or 3.0?

To avoid this, the smart city must be implemented in a spirit of contribution and co-design between the various stakeholders, including citizens who would take ownership of the concept as part of co-management with the municipality and private actors. However, the technologies, and in particular the AI associated with smart cities, are the domain of specialists, with knowledge of solutions held by companies. The challenge for the inhabitants of these new cities is to regain control of their own technology from suppliers. And to do this, they need to understand and be involved in it. Cities such as Dublin[3] and Madrid[4] have developed such initiatives.

6.4. Methodology and results

On the basis of our semi-structured interviews, we have identified four main pathways which, if taken into account, could lead to the development of a smart city that is not solely technophile, and in which businesses will play an important role. These pathways are elucidated through excerpts from interviews in the footnotes.

The *first pathway* is to make cities, urban infrastructures and public services more efficient. From this point of view, the use of AI and data is the preamble to the concept of smart cities. Digitization should ideally help to improve local public policies, at an expected lower cost. This implies the following courses of action:

– Identifing the costs and opportunities of smart cities:

- We need to redefine the measurement of public performance by mobilizing all available expertise (university research, public bodies, think tanks). We will then need to draw on this expertise to develop local public

3 With the initiative of its city council and its "Beta Projects" project, the city of Dublin has put in place a framework facilitating experimentation and its evaluation by involving local residents.

4 Madrid had the maturity and wisdom to allow the "Campo de Cebada" process to unfold under the leadership of citizens who became involved in the future of the La Latina district after rehabilitation plans with the private sector came to a halt following the crisis of 2009.

policies, in order to locate and eliminate duplication made unnecessary by ICTs.

– Boosting government performance through digital technology and AI in particular:

- the performance of public services requires the development of e-administration, which can take the form of a one-stop digital platform. It will also be necessary to set up a digital smart cities community to exchange best practices, coupled with the equivalent of an urban needs observatory.

– Improving the quality of life and making the city more attractive (Clétil 2018):

- the aim is to facilitate access to public and private services by better adapting them to employees' work rhythms and schedules and by developing connected supports using high-performance AI (widespread Wi-Fi, digital signs, QR codes) while combating crime and incivility[5].

– Promoting greater mobility:

- with a view to optimizing travel and combating urban congestion, we can imagine, for example, AIs that make it easier for users to get around, such as those already available on smartphones. It could also mean making surplus transport and parking capacity available to other public or private organizations or setting up "coworking" spaces[6].

– Developing environmental skills:

- the aim is to collectively reduce the ecological footprint by optimizing consumption at the level of a city or even a metropolis (smart grids)[7]. This implies greater autonomy for cities to organize their consumption patterns and to test new energy models or other innovative devices within the framework of sustainable development, all within a balanced budget for the

---

5 "The interest of data is not only to inform and promote services to the public, but also to create new ones, such as connected walking sticks for the visually impaired and blind", says Jean-Marc Lazaud, CEO of Open Data Soft (authors' translation).

6 "I'm thinking of a connected territory where we pay attention to ecomobility, respect for the environment, etc.," David Lacombled, president of the Villa Numeris think tank (authors' translation).

7 "In 2017, the Alpes-Maritimes departmental council launched the smart deal with businesses, this being linked to sustainable development," Charles-Ange Ginésy, president of the Alpes-Maritimes departmental council (authors' translation).

city, following the example of the Nice Côte d'Azur metropolis, which is part of the collaborative European IRIS project[8].

The development of new public–private partnerships to stimulate economic spin-offs corresponds to the *second pathway*. In this context, the concept of smart cities implies testing new partnership models between municipalities and businesses, to better respond to the challenges facing cities, and also contribute to the development of innovative businesses. These partnerships must achieve the following:

– Identify an appropriate business model[9]: this involves implementing a Local Small Business Act for the awarding of tenders. It must also attract and energize human capital, as well as foster collaboration between large companies and start-ups by setting up coworking spaces and incubators, new places for disseminating and sharing knowledge.

– Enhance the value of smart city companies, particularly for citizens, in the day-to-day running of the city[10].

The *third pathway* is to involve citizens in the design and development of the smart city[11]. Today's urban challenges no longer seem to be solved by top-down intervention alone, but increasingly by the ingenuity of local residents. Here, we need to optimize participatory approaches with citizens and companies. With this in mind, we need to develop direct relations between elected representatives, on the one hand, and businesses and citizens, on the other hand. By way of illustration, we can imagine a collaborative platform enabling city dwellers to get involved in monitoring and financing projects (crowdfunding), and in public service missions, so

8 The IRIS (Integrated and Replicable Solutions for Co-Creation in Sustainable Cities) project, involving 43 European partners, consists of energy management through innovative approaches in the energy, transport and ICT sectors in three major European smart cities (Nice, Utrecht and Gothenburg). The Nice Côte d'Azur metropolis will receive a grant of 4.3 million euros for this project.

9 "We can envisage a public-private rebalancing through data, as, for example, in a data-for-services type of logic," Régis Chatelier, Director of the LING (authors' translation).

10 "The city of the future is built on an ecosystem of companies that will be there to provide services," says Patrice Slupowski, Director of Innovation for Orange Business Services (authors' translation).

11 "François Veauleger, Director of the Montgenèvre Tourist Office, explains: "It is vital to integrate user behavior on site and involve them in the development of a smart logic" (authors' translation).

that this smart city is centered on the human rather than the technical. AI will play an important role here.

The *fourth and final pathway* deals with the implementation of open data at the level of administration, residents and businesses. At this level, it is clearly recognized that open data combined with AI can strengthen democratic practices and generate economic spin-offs within a local authority, as well as at the level of any organization[12]. We identify the following two points:

– Develop an open data ecosystem: the aim is to put open data at the service of users and businesses alike[13], by simplifying data so that it is presented in an understandable way, through a joint effort by the municipality and developers. This open environment involves the exchange and sharing of data between economic stakeholders and public actors[14].

– Protecting businesses and citizens, as well as other smart city stakeholders, from digital risks: elected municipal officials will have to develop actions to reduce the digital divide between citizens, and also between businesses, to effectively train municipal staff and elected officials in ICT, to guard against potential cyber-attacks, and to reinforce the security of the smart city's digital networks. This requires constant vigilance during network installation and maintenance[15].

---

12 "The law requires local authorities with more than 3,500 inhabitants and 50 employees to implement an open data policy. However, of the 4,000 communes required to comply by October 2018, only 250 have done so", Cédric Verreaux (op. cit., authors' translation).
13 "For this data to be truly useful, cities and businesses alike will need to develop interoperable, open and secure infrastructures to transmit and share this wealth of information. This is what will enable the development of tomorrow's smart city enterprise. Why not imagine, in reception halls, next to screens displaying company news, the state of road traffic, the waiting line at the company restaurant, or at local shops right up to the dry cleaner's, not forgetting the CoWorking Space in which the company has places for its employees?", Louis Treussard, CEO BNP Paribas (authors' translation).
14 "Data is becoming a new local or regional public policy," says Luc Belot, Managing Director of the Réalités group's Hub 5 subsidiary (authors' translation).
15 "Security is one of the pillars of sustainability for our future smart cities," Nicolas Chalvin, marketing director at IoT Solutions & Services Gemalto.

6.5. Conclusion

The use of technology and digital technology enables a city to optimize its flows and thus make them more efficient and productive, under the aegis of centralized urban governance and with the support of major corporations. This is currently the trend in smart city concepts, which focus on technical devices and fail to address the city's social and cultural issues.

However, as Adam Greenfield (2013) points out, to avoid the smart city being in the hands of datacrats, all citizens worldwide need to think carefully and collectively about the place they want to give to the various technologies associated with the running of cities, and in particular to ICT, and what these technologies can do for them in terms of meeting their expectations. What's more, it is the responsibility of all those who have some basic knowledge of networked digital information technologies to raise awareness, explain and contextualize them to all individuals, so that it is easy to make the decisions needed to create a city that puts the human dimension at its heart.

Le cube, which is a digital creation center[16], talks about an "agile city" developing a model based on decentralized, bottom-up innovations, integrating AI and involving citizens in the co-construction of their city in close collaboration with companies. It identifies three areas of development:

– The "Lego" city is based on the co-construction of open urban systems that can be improved by their inhabitants. The aim is to raise awareness of co-design and ICT among businesses and citizens and to develop "Open Source" projects[17].

– The "local city" encourages the development of small-scale projects and local initiatives by citizens and businesses, trying to create a pollination effect. The aim is also to map initiatives and resources in order to create a larger-scale network.

– The "lab city" draws on the creativity, personality, knowledge and resources of citizens and the entrepreneurial fabric to produce innovation. This can be seen in the creation of new public spaces (Makers Labs, FabLabs, Repair Cafés, 3D printer bars) and the widespread use of crowdfunding initiatives in which users and entrepreneurs are stakeholders.

16 Available at: www.lecube.com.

17 The Open Source explained in Lego, see: http://youtu.be/a8fHgx9mE5U.

All this thinking tends to show that the future of smart cities must be based on a bottom-up vision that aims to integrate the company and the citizen as actor and "shaper" of their city (Vanin and Cyrulnik 2019). In this context, reflection on the ethical uses of AI is essential.

6.6. References

Barnier, V. and Tucoulet, C. (eds) (1999). *Ville et environnement : de l'écologie urbaine à la ville durable*. La Documentation française, Paris.

Beretta, J. (2015). Les caractéristiques d'une ville intelligente. Dossier Smart Grids de la CRE (Commission de Régulation de l'énergie).

Brasseur, B. and Biaz, F. (2018). L'impact de la digitalisation des organisations sur le rapport au travail : entre aliénation et émancipation. *Question de management*, 21, 143–155.

Brynjolfsonn, E. and McAfee, A. (2011). *Race Against the Machine: How the Digital Revolution is Accelerating Innovation, Driving Productivity, and Irreversiby Transforming Employement and the Economy*. Digital Press, Cambridge.

Calzada, I. and Satyam, A. (2017). *The Smart City Transformations*, 1st edition. Bloomsbury, London.

Caragliu, A., Del Bo, C., Nijkamp, P. (2009). Smart cities in Europe. In *Creating Smarter Cities Conference*. Edinburgh Napier University, Edinburgh.

Castells, M. (1996). *Rise of the Network Society: The Information Age*. Blackwell, Cambridge.

Cazals, F. (2015). *Stratégies digitales : la méthode des 6C*. De Boeck supérieur, Paris.

Cletil, G. (2018). Smart City, innover avec précaution. *L'Avenir Côte d'Azur*, 2400, 10–13.

Coëffé, T. (2017). Baromètre du numérique 2017 : équipement, usages et compétences numériques des Français. *Blog du modérateur*.

Dorn, D. (2017). La montée en puissance des machines : comment l'ordinateur a changé le travail. *Revue française des affaires sociales*, 1, 35–63.

Dutton, W.H. (1997). *Wired Cities: Shaping the Future of Communications*. Macmillan, London.

El Hadi, M. and Arsenault, C. (2011). Introduction. Dynamisme et stabilité dans l'organisation des connaissances : d'une conférence à l'autre, Toronto 2000. In *L'organisation des connaissances : dynamisme et stabilité*, El Hadi, M. (ed.). Hermès-Lavoisier, Paris.

Eveno, E. (2018). La Ville intelligente : objet au cœur de nombreuses controverses. *Quaderni*, 96(2), 29–41.

Flichy, P. (2011). Le sacre de l'amateur. Sociologie des passions ordinaires à l'ère numérique. *Communication et Organisation*, 39, 157–259.

Gassman, O., Bohm, J., Palmie, M. (2019). *Smart Cities: Introducing Digital Innovation to Cities*. Emerald Publishing Limited, London.

Giffenger, R., Fertner, C., Kramar, H., Kalasek, R., Pichler-Milanovic, N., Meijers, E. (2007). Smart cities – Ranking of European medium-sized cities. Centre of Regional Science, Vienna [Online]. Available at: http://www.smart-cities.eu [Accessed 25 November 2010].

Greenfield, A. (2013). *Against the Smart Cities*. Do Project, New York.

Guarrigues, A. and Blanc, S. (2016). La Smart city à la recherche de modèles économique. Club Techni.Cité [Online]. Available at: http://www.lagazettedescommunes.com/441803/la-smart-city-a-la-recherche-de-modeles-economiques/#sommaireDossierTarget.

Guillau, H. (2015). Une nouvelle conception des services publics numériques inspirée par le gouvernement britannique. *Le Monde.fr, M Blog, le blog d'Hubert Guillau, de Xavier De La Porte et de Remi Sussan*.

Haentjens, J. (2021). *Smart city, ville intelligente: quels modèles pour demain ?* La Documentation française, Paris.

Hollands, R. (2008). Will the real smart city please stand up? *City*, 12(3), 303–320.

Ingallina, P. (2019). *Ecocity, Knowledge city, Smart city : vers une ville écosoutenable ?*, 1st edition. Septentrion, Paris.

Jollivet, M. (2001). *Le Développement durable, de l'utopie au concept : de nouveaux chantiers pour la recherche*. Elsevier, Paris.

Kominos, N., Pallot, M., Schaffers, H. (2013). Special issue on smart cities and the future Internet in Europe. *Journal of the Knowledge Economy*, 4(2), 431–446.

Kourtit, K. and Nijkamp, P. (2012). Smart cities in the innovation age, innovation. *The European Journal of Social Science Research*, 25(2), 93–95.

Languillon-Aussel, R. (2018). Le programme " smart communities " au Japon. Nouveaux enjeux de pouvoir des ressources et des systèmes d'information urbains. *Flux*, 114(4), 38–55.

Leea, J-H., Phaalb, R., Lee, S-H. (2013). An integrated service-device-technology roadmap for smart city development. *Technological Forecasting and Social Change*, 80(2), 286–306.

Mahinzan, A. (1999). Smart cities. The Singapore case. *Cities*, 16(1), 13–18.

Moreno, C. (2013). Les défis de la smart city humaine. In *Colloque Smart City Exhibition 2013. Communication, Quality and Development in Smart Cities*, Bologna.

Picon, A. (2015). *Smart Cities. Théorie et critique d'un idéal auto-réalisateur*. B2, Paris.

Schneider Electric (2014). La ville de demain : les piliers de la Smart City [Online]. Available at: http://blog.schneider-electric.fr/gestion-de-lenergie/2014/03/25/les-piliers-smart-city/.

Simondon, G. (2012). *Du mode d'existence des objets techniques*. Aubier, Paris.

Townsend, A. (2014). *Smart Cities: Big Data, Civic Hackers, and the Quest for a New Utopia*, 1st edition. W.W. Norton & Company, London.

Vanin, L. and Cyrulnik, B. (2019). *Smart City, une autre lecture de la ville*. Ovadia, Paris.

Wiener, N. (2014). *La cybernétique : information et régulation dans le vivant et la machine* (translated posthumously by Le Roux, R., Vallee, R., Vallee-Levy, N.). Le Seuil, Paris.

Wolframm, M. (2012). Deconstructing smart cities: An intertextual reading of concepts and practices for integrated urban and ICT development. In *P. REALCORP Proceedings*. CEIT Alanova, Vienna, 171–181.

PART 3

AI at the Service of the Community: A World to Build Towards

7

Artificial Intelligence: A New Challenge for Information and Media Literacy

7.1. Introduction

Changes in the world of information and media, under the influence of digital technology, are numerous, well known and widely documented.

The acceleration of the media *tempo*, the increase in the flow of information, competition between media, and the renewal of formats and media for disseminating information are all changes that have shaken the professional practices of journalists and the informational practices of audiences.

Artificial intelligence, the latest development in information and communication technology (ICT), is profoundly changing the information ecosystem and is currently causing concern by offering new tools for producing and distributing information. This, coupled with the strength of social media, which capture a hegemonic share of online information traffic, is profoundly altering the way in which individuals communicate, inform themselves, and also sometimes misinform each other.

Far from being a novelty, artificial intelligence has been part of the media landscape for many years. Although it has been the subject of fantasies, particularly in science fiction literature and cinema, it is only very recently that it has aroused a great deal of interest, particularly because it takes diverse, concrete and, above all, accessible forms.

Chapter written by Loïc SUMIEN.

Following the example of ChatGPT, BARD, BingChat and Midjourney, which burst onto the scene between late 2022 and early 2023, artificial intelligence is opening up "a new ecosystem" (Portnoff and Soupizet 2018), raising questions about the ethical use and learning process of this technology, which calls for the definition of the best possible relationship between man and machine.

For the first time, this technology is raising questions about the future of many so-called "intellectual" creative professions in particular, following the example of Hollywood screenwriters who, after long months on strike, have obtained a status quo on the subject. Where the first intelligent machines, which were not really intelligent at all, were confined to replacing first-level professions, which until then had caused little stir.

Artificial intelligence now regularly fuels public debate, even on a daily basis, about its alleged power to do harm. These technologies are developing at breakneck speed, in a short space of time as opposed to the long time required for training, education or regulation. It is therefore imperative to measure one's capacity to do harm as well as good so as not to let oneself be impressed by a chimera that could ultimately be an opportunity.

Beyond the initial panic effect, artificial intelligence (AI) undeniably presents numerous potentialities, but like all technologies before it, it is not without its share of worries, mobilizing many researchers who are calling for a moratorium on the subject, even calling on the giants of the tech industry, such as the charismatic boss of Tesla, SpaceX, Neuralink and X, Elon Musk, co-signatory of the letter Pause Giant AI Experiments[1].

To meet the challenges raised by AI, a number of actors need to be mobilized, including information and media literacy (IML). While this has long been confined to a "technical" approach to the media – to which we shall return to later – it is now a question of encouraging new pedagogical practices, particularly in conjunction with the information and communication sciences.

From now on, IML will be a key factor in the support, growth and learning of new media, and here, in the context of the deployment of AI, in order to acquire an ethical, reasoned and enlightened use of it. However,

1 See: https://futureoflife.org/wp-content/uploads/2023/05/FLI_Pause-Giant-AI-Experiments_An-Open-Letter.pdf.

while AI is often seen as a miracle cure for society's problems with new media, we shall see that it must continue to reinvent itself if it is to live up to the ambitions it is expected to meet.

7.2. IML: challenges, issues and perspectives

The first public policies on information and media literacy were discussed at *symposia*, congresses, conferences and other colloquia, including Grünwald in 1982, Alexandria in 2005 and Paris in 2007. IML has been a focus of attention since the 1980s, but it was not until the early 2000s that it came into its own as a response to the many challenges posed by the exponential development of information and media environments in the digital spectrum.

Since then, IML has been the subject of opinions (CESE 2019)[2] and even proactive public policies, or at the very least has been mentioned in public policy speeches[3] (Corroy and Froissart 2018), which encourage its deployment in and out of school, so that it takes on a more important role in the educational sphere in particular, by encouraging teachers to integrate this learning transversally into their teaching.

While the analysis of the written press and traditional media used to be the mainstay of public IML policies, new issues are now shaping public IML policies, such as the fight against proselytizing, the proliferation of online disinformation and, more recently, the development of AI. Media and information literacy appears to be one of the levers to be mobilized, to help individuals of all ages to appropriate these new environments and their actors.

Initiated by the work of UNESCO, public policies for IML in France are steered by the *Centre pour l'éducation aux médias et à l'information* (CLEMI). This National Education body is present in every *académie* in France, under the authority of the *recteur* and aims to develop IML in France, specifically within the school system, although it may occasionally intervene outside the school.

2 Available at: www.lecese.fr/sites/default/files/pdf/Avis/2019/2019_30_defi_education_medias_information.pdf.

3 Corroy, L. and Froissart, P. (2018). L'éducation aux médias dans les discours des ministres de l'Éducation (2005–2017). *Questions de communication*, 34.

Long neglected, even considered marginal, IML became, on paper, a priority for the French Ministry of Education from 2015 onwards with *"Le plan numérique pour l'éducation"*. The attacks that have severely affected France in recent years, and more recently the murders of teachers Samuel Paty and Dominique Bernard, have accelerated the desire to put in place genuine public media literacy policies, a possible last line of defense to contain cyberphenomena that are gangrenous on the web: "Following the attacks of January 2015, IML was reinforced with the objectives of citizenship education and the transmission at school, of a culture of the press and freedom of expression"[4] (author's translation). This awareness has been voiced by those in charge of public education policies, in the face of the many issues and challenges posed by the evolution of digital information media.

7.2.1. *The role of IML in the development of AI*

More than just a topical issue, AI is a real challenge for society and falls squarely within the remit of information and media literacy. With a notable irruption at the heart of the information ecosystem, AI represents a global issue that is attracting strong involvement from public and private actors, including professionals in the information and media sectors, as well as numerous researchers.

While AI is not a new technology, over the past few months, the term has become a permanent fixture in the media debate. Two key events appear to be at the root of this sudden popularity: the launch of ChatGPT in December 2022 and again in early 2023, and the launch of two new competing conversational agents, BARD and BingChat, the latter officially launching the AI race contested by the giants GAFAM (or GAMMA since Facebook is called Meta).

Economics, politics and education – all these fields seem to be affected in one way or another by AI, raising many questions about its power to act and, above all, its control.

Already widely deployed in the French education system, IML has the onerous task of educating students to be good media citizens and, at the

4 See: https://www.clemi.fr/fileadmin/user_upload/guide_webradio/EMI_et_strategie_numerique_cycle_4_546409.pdf.

same time, encouraging them to cultivate their critical faculties, a skill described as indispensable for evolving in the complex worlds of information 2.0.

Through information and media literacy (IML), students learn to become responsible citizens in a society marked by the multiplication and acceleration of information flows. They develop critical thinking skills and are able to act in an informed way to seek out, receive, produce and disseminate information via increasingly diversified media[5] (author's translation).

Knowledge and skills that are the subject of a common core are taught, in theory, across the board from elementary school through to university. We say in theory, because its uniform application, sometimes contested, remains at the discretion of each school and is the subject of debate as to its effective deployment in the educational field.

Initially focused on the study and analysis of the printed press and early mass media, the digital revolutions of the last two decades have accelerated a shift in pedagogical paradigm, with teaching now geared towards the proper use of digital media, Internet 2.0 and, more specifically, digital social media.

More recently, in response to current phenomena such as the successive crises brought about by Covid-19, IML is trying to take its place in the fight against the spread of false information online and fake news, which are themselves amplified by AI.

IML is now more positivist than critical, as was long the case when faced with the emergence of new technologies whose massive use worried schools, which for a long time remained impervious to these external phenomena.

AI must fit into this new paradigm, with the learning of an open and enlightened vision of the potential offered by the various forms of AI, which require prerequisites for responsible use.

7.2.2. *IML for an ethical, reasoned and enlightened use of AI*

In line with existing IML programs, the aim is to provide students with the keys to reading and understanding this phenomenon on several levels. A

5 See: https://eduscol.education.fr/1531/education-aux-medias-et-l-information.

multiscalar vision that will enable them to understand the workings and mechanisms of this phenomenon, and thus learn how to extract the very essence of it, so as not to allow themselves to be possessed by this tool, which must remain at the service of mankind. The aim is to strengthen information and media literacy in the light of the new questions raised by AI in terms of human/machine relations and also in terms of its role as an intermediary between the transmission and reception of information. While the skills already identified in IML for dealing with the media environment are a solid foundation, the challenge now is not only to reinforce these initiatives, but also to create new ones:

> Take institutional actions to enhance AI literacy across all sectors of society: Provide basic AI education to all citizens, educating them on thinking critically and responsibly about their choices, rights and privileges in the context of AI and its impact on their day-to-day lives. Inform them on how to protect their privacy and control their own data and decisions. Dismantle the myths and hype around AI by educating the population about its limitations, as well as the differences between AI and human intelligence. Carefully integrate emerging AI literacy skills with existing foundational skills such as information and media literacies, and identify ways of merging the different required literacies to prevent overloading the curricula[6] (UNESCO 2021, author's translation).

A worldwide realization by international organizations of the importance of IML in accompanying the development of AI, recently highlighted by Sylvie Retailleau, Minister of Higher Education, in an interview with Le Parisien dated September 11, 2023. The latter "calls on the educational community to train pupils and students to make better use of the powerful tools of generative AI" and warns of a risk of "a generational gap in the use of a Chatbot like ChatGPT"[7] (author's translation).

For a long time, media literacy has been confined to learning through the media and has not been devoted to developing knowledge of media and

6 See: https://teachertaskforce.org/sites/default/files/2023-07/2021_UNESCO_AI-and-education-Guidande-for-policy-makers_EN.pdf.

7 Available at: www-leparisien-fr.cdn.ampproject.org/c/s/www.leparisien.fr/amp/high-tech/sylvie-retailleau-il-faut-apprendre-aux-eleves-a-bien-utiliser-lia-pas-linterdire-11-09-2023-7F3ZIQJ5EVC75H3OL4LCXPUPLQ.php.

information ecosystems, with real reflection on the media (their environment, their economy, the way they operate, etc.). Learning about the media through digital tools requires new pedagogical methods for acquiring an ethical and reasoned use of them.

Over and above the regulatory measures currently being implemented by the public authorities, and given the very nature of the phenomenon, which is now a burning issue, the misuse of AI represents a challenge for the future. It is naturally up to IML to take up this issue in the educational field, by making available a repository of best practices in the use of AI, as well as enabling the acquisition of skills so that everyone can grasp these new tools with their exponential potential.

Often in reaction, IML, which is already trying to adapt to the various transformations in information and media environments, is once again faced with a renewal of its practices, the rapid evolution of this technology being naturally at odds with the long timeframe of training and its implementation in the educational field.

This education in AI has been the subject of reflection and recommendations since 2021 by UNESCO, set out in a document entitled "AI and education: a guide for policy-makers":

> Integrate AI-related skills into school curricula and technical and vocational education and training (TVET) qualifications: Enact changes in curricula to prepare students for the future, ensuring their relevance to shifting economies, labour markets and societies across all subjects and competencies. Develop courses, programmes and qualifications to provide awareness and expertise around how AI technologies work, their ethical implications, and how they should be designed. Support the development of tools for learning about AI that are underpinned by pedagogical research and sound methodologies[8] (UNESCO 2021).

While these various measures and recommendations have the merit of existing, they could nonetheless be based more on the lessons already taught in IML, since they take up common notions such as critical thinking, and

8 See: https://teachertaskforce.org/sites/default/files/2023-07/2021_UNESCO_AI-and-education-Guidande-for-policy-makers_EN.pdf.

thus avoid setting up yet another "education for" or "overloading the programs"[9].

7.2.3. *AI and misinformation: critical thinking as the backbone of IML in France*

The responsible use of AI calls for critical thinking, a notion that is widely studied in IML. AI certainly goes one step further in the fabrication and dissemination of false information online. This is an issue that is already of concern to IML and is the subject of specific recommendations and workshops. While the medium is new, there is a repository of knowledge and skills to be mobilized, including digital literacy, information literacy and critical thinking.

Like social networks before it, AI blurs the trail and makes it even more difficult to identify the source of information. It calls into question the receiver's ability to analyze its origin, and thereby considerably complicates its interpretation.

Since 2016 at least, the rise of online disinformation has called for the development of public policies that encourage the amplification of concrete actions to curb or at least circumscribe this phenomenon in order to protect the population from mass online disinformation. Numerous initiatives have been guided by a common ambition, to make students enlightened citizens in information and media ecosystems to "develop a civic practice of the media by having a critical and distanced reading of content" (CESE 2019, author's translation).

We are entering "a new era of falsehood", as William Andureau[10] (author's translation), a journalist at the Le Monde newspaper, in charge of the fight against false information and the study of conspiracy in the "Les Décodeurs" column, puts it.

In an article dated April 13, 2023, on the subject of the Midjourney application, he highlights the complexity of identifying the true from the false, even for seasoned information professionals. A new era of misinformation, at a time when the codes of the "old" new media have not

9 *Ibid.*

10 Available at: www.lemonde.fr/idees/article/2023/04/13/les-images-concues-par-intelligence-artificielle-nouvel-outil-de-desinformation_6169318_3232.html.

been mastered by all, and when international news floods the networks with thousands of false reports every day.

So what is the role of media literacy? First and foremost, it is undoubtedly to encourage individuals to acquire a critical mindset by promoting:

> The "media literacy" program is designed to enable students to know, read, understand and appreciate the representations and messages of the different types of media with which they are confronted on a daily basis, to find their way around them and to use these major media and the content they convey in a relevant, critical and thoughtful manner[11].

It also means fostering a culture of access to information that:

> presupposes an autonomous, critical and creative use of information, going as far as the production of knowledge, and "information culture" which extends to media literacy, ethical considerations and presupposes "social integration"[12] (Juanals 2003, author's translation).

This is what defines an individual's ability to access and analyze information before relaying it:

> The fact of having a set of knowledge and skills enabling one to recognize the existence of an information need and to have the ability and discipline to find the appropriate information, understand it, exploit it and communicate it, in a problem-solving or decision-making perspective[13] (author's translation).

IML must strive to encourage the use of virtuous AI, by training in the use of AI that can help the receiver identify false information in the image of

---

11 See: https://cache.media.eduscol.education.fr/file/EMI/41/1/L_EMI_en_10_definitions_546411.pdf.

12 Simonnot, B. (2009). Culture informationnelle, culture numérique : au-delà de l'utilitaire. *Les Cahiers du numérique*, 3(5), 25–37 [Online]. Available at: www.cairn.info/revue-les-cahiers-du-numerique-2009-3-page-25.htm.

13 See: https://vitrinelinguistique.oqlf.gouv.qc.ca/fiche-gdt/fiche/8363208/culture-de-linformation.

"DeFaktS" or "noFake" – two tools that put the often pessimistic rhetoric about AI at some distance.

7.3. AI and media literacy: taking a systems approach

While IML is increasingly in demand, it still raises many questions about its application and practical integration in the educational sphere. A number of obstacles have been identified by the CESE[14] in 2019, including a lack of synergy between actors, schemes that are often unfamiliar to the general public, a lack of training for teachers and information professionals in IML and a lack of resources allocated to the deployment of public policies.

Of course, in principle, the massive development of IML can only be encouraged, but for the time being, it alone cannot answer all the questions raised by the emergence of AI. A whole universe needs to be built to meet the many challenges raised by the exponential development of AI. In order to provide the best possible support for our audiences, we believe that IML should be part of a more systemic approach (Degenne 1977).

The media, education, institutional organizations and associations all have a role to play in raising public awareness, as it is the coordinated actions of these actors that will best meet the many challenges posed. The aim is to take account of a complex media environment, changing audience information practices, and territorial and social inequalities.

So, although IML has a fundamental role to play, we cannot say that it is enough as it stands. It is facing a renewal of its practices and will have to strive to reduce the gap between technological time and training time. What is more, given that the latter is linked to the socio-political context, to changes in the information and media sphere, and even to the theories that punctuate the information and communication sciences, IML is called upon to reinvent itself in response to this new challenge.

7.4. AI, a necessary consideration in the educational field

Beyond IML , AI is going to take over the educational field, like the Internet and the social web before it. In a recent publication, the United

14 Available at: www.lecese.fr/sites/default/files/pdf/Avis/2019/2019_30_defi_education_medi-as_information.pdf.

Nations Educational, Scientific and Cultural Organization (UNESCO) estimates that AI will be worth $6 billion in education by 2024[15] (UNESCO 2021).

Schools need to open up to new uses, and even new teaching methods, by making the most of the potential offered by AI, in the context of learning. If it has already appeared in the background in recent years:

> Evidence remains scarce on how AI can improve learning outcomes and whether it can help scientists and education professionals better understand how effective learning occurs[16] (Zawacki-Richter et al. 2019, author's translation).

Initially impervious to technological phenomena in an approach we describe as protectionist (Corroy-Labardens et al. 2015; Corroy-Labardens 2022), schools must now make these devices their own in order to train tomorrow's citizens in useful, current and, above all, future knowledge.

The educational sphere, by seizing AI, must propose new ways of learning by summoning this "acculturation to the digital" (Cerisier 2011) that has already been talked about for many years with the aim of "enabling everyone to understand how AI mechanisms work, in order to develop a critical and creative eye in relation to its current and future uses"[17] (Heiser and Romero 2023, author's translation).

This education in AI, which could find its place within the framework of IML, should enable all publics, and of course the youngest, to "acquire and develop a solid understanding of AI", enabling them, among other things, to reduce inequalities and a potential future divide that would feed pre-existing gaps in the most fragile publics (which does not necessarily mean the youngest).

Against this backdrop, UNESCO is leading the way, defining four objectives to be achieved over the coming years[18]:

> ensure the inclusive and equitable use of AI in education; make the most of AI to improve education and learning; promote the

15 See: https://unesdoc.unesco.org/ark:/48223/pf0000380006.
16 *Ibid.*
17 See: https://hal.science/hal-04013223v2.
18 See: https://unesdoc.unesco.

development of skills needed for life in the AI age, including teaching how AI works and its implications for humanity; ensure the transparent and auditable use of education data.

IML is undergoing a transformation, and its various educational and training programs are geared towards learning about and becoming acculturated to the new digital tools and phenomena so that everyone can seize the potential offered by the latter. In this sense, AI will undoubtedly be one of the greatest challenges to be met in the coming decades, both in education and in the information sphere, in order to strike the right balance between man and machine.

Aware of the need for change, national and international bodies are now directing their training and awareness-raising initiatives in this direction, particularly in contexts of economic, social, political, democratic or health crisis.

Constantly changing, the information and media ecosystem needs to be closely monitored, in order to provide adequate responses to contemporary problems, transforming them into opportunities and making them accessible to everyone.

7.5. References

Andureau, W. (2023). Les images conçues par intelligence artificielle, nouvel outil de désinformation. *Le Monde*.

Bengio, Y., Wozniak, S., Musk, E., Noah Harari, Y. et al. (2023). Pause Giant AI Experiments: An Open Letter. UQAM, Montréal.

Cerisier, J.-F. (2011). Acculturation numérique et médiation instrumentale. Le cas des adolescents français. Thesis, Sciences de l'information et de la communication. Université de Poitiers, Poitiers.

Collin, S. and Marceau, E. (2021). L'intelligence artificielle en éducation : enjeux de justice. *Formation-profession*, 29(1). doi: 10.18162/fp.2021.a230.

Corroy, L. and Froissart, P. (2018). L'éducation aux médias dans les discours des ministres de l'Éducation (2005-2017). *Questions de communication*, 173–188. doi: 10.4000/questionsdecommunication.15749.

Corroy-Labardens, L. (2022). *Éducation aux médias en Europe. Histoire, enjeux et perspectives*. L'Harmattan, Paris.

Corroy-Labardens, L., Barbey Weabey, F., Kiyindou, A. (eds) (2015). *Éducation aux médias à l'heure des réseaux, Colloques & rencontres*. L'Harmattan, Paris.

De Cazals, M. (2010). La saisine du Conseil économique, social et environnemental par voie de pétition citoyenne : gage d'une Ve République " plus démocratique " ? *Revue française de droit constitutionnel*, 82, 289–312.

Degenne, A. (1978). Review of: Le Moigne Jean-Louis, La théorie du système général. Théorie de la modélisation. *Revue française de sociologie*, 19(2), 300–301.

Heiser, L. and Romero, M. (2023). Education à l'intelligence artificielle : quelles compétences acquérir par les élèves ? Report, Université Côte d'Azur.

Juanals, B. (2003). *La culture de l'information : du livre au numérique*. Hermès-Lavoisier, Paris.

Lecas, G. (2006). La modélisation systémique : outils méthodologiques pour économistes. *Innovations*, 24, 199. doi: 10.3917/inno.024.0199.

Licata Caruso, D. (2023). Sylvie Retailleau : "Il faut apprendre aux élèves à bien utiliser l'IA, pas l'interdire". *Le Parisien*.

Margarida, R. and Heiser, L. (n.d.). Enseigner et apprendre à l'ère de l'intelligence artificielle. White Paper, Canopé.

Miao, F., Holmes, W., Ronghuai, H., Hui, Z. (2021). *IA et éducation : guide pour les décideurs politiques*. UNESCO, Paris.

Morin, E. (2008). *La méthode*. Le Seuil, Paris.

Nauroy, D. (2004). Brigitte Juanals, La culture de l'information. Du livre au numérique. *Question de communication*, 374–376. doi: 10.4000/questionsdecommunication.5915.

Papy, F. (2008). *Problématiques émergentes dans les sciences de l'information, traité des sciences et techniques de l'information*. Hermès-Lavoisier, Paris.

Pereira, I. (2018). Les grammaires de l'éducation critique aux médias à l'épreuve du numérique. In *Éducation aux médias à l'heure des réseaux*, Corroy-Labardens, L., Barbey, F., Kiyindou, A. (eds). L'Harmattan, Paris. doi: 10.4000/communication.7452.

Portnoff, A.-Y. and Soupizet, J.-F. (2018). Intelligence artificielle : opportunités et risques. *Futuribles*, 426, 5. doi: 10.3917/futur.426.0005.

Simonnot, B. (2009). Culture informationnelle, culture numérique : au-delà de l'utilitaire. *Les Cahiers du Numérique*, 5, 25–37.

Zawacki-Richter, O., Marín, V.I., Bond, M., Gouverneur, F. (2019). Systematic review of research on artificial intelligence applications in higher education – where are the educators? *International Journal of Educational Technology in Higher Education*, 16, 39. doi: 10.1186/s41239-019-0171-0.

8

Agenda-Setting AI Algorithms, What is the Agenda?

8.1. Introduction

Since their development in the early 2010s, artificial intelligence (AI) algorithms have played a key role in the selection and presentation of content on social networks, profoundly influencing user interactions and decisions. Introduced without warning into our phones, these algorithms now act as active mediators in digital communication, evolving to sort and interpret information in the manner of news editors. The content fostered by these systems shapes an immersive digital reality, particularly striking among young users. For example, TikTok uses sophisticated algorithms to manipulate and filter the content presented, stimulating engagement while introducing significant biases.

The methodology of this study begins with the creation of an AI algorithm simulator inspired by TikTok's recommendation system. Using data on user interactions and video metadata, an AI model was developed with Python libraries such as Scikit-learn and TensorFlow. A user interface was implemented via Flask to facilitate interactions. The results highlighted several principles of "algorithmic thinking", including reinforcement based on user feedback, optimization of specific goals and dependence on historical data.

Next, node analysis was used to examine the actions of influencers in correlation with the algorithms. This methodological approach explores the

Chapter written by Horea Mihai BADAU.

relationships between entities, identifying influencer accounts as nodes and their interactions as links. Data collected via the TikTok API was analyzed using NodeXL, revealing the most dominant influencers potentially influenced by the algorithms.

The third stage of the study focuses on identifying semantic universes on TikTok by using the platform's API to collect and analyze influencer publications. Following a similar methodology to an earlier study on Facebook and Instagram, this phase seeks to examine the impact of AI algorithms on users' semantic universes, exploring the recurring patterns and themes present in their interactions.

Finally, the fourth stage aims to identify "rabbit holes" within TikTok's semantic universes, resulting from Douyin's collaborative filtering. A study in Sweden revealed that the motivations of algorithm-influenced users include entertainment, information seeking, leisure, social status and discovering new interests. However, many users actively try to avoid certain types of content, highlighting the challenges of involuntary exposure to personalized content suggested by the algorithm.

The study concludes that continuous exposure to TikTok's algorithms pushes influencers to internalize algorithmic logic, modifying their behavior to maximize engagement at the expense of content diversity and quality. This adaptation engenders a vicious circle in which popularity reinforces algorithmic visibility, impoverishing users' semantic universes and accentuating the effects of less-regulated "rabbit holes".

8.2. Research questions

Influencers spend a lot of time interacting with AI algorithms, and their feedback is vital to their success. Can the actions of AI algorithms, such as the choice of information they select and publish on users' walls, have an effect similar to the agenda effect of TV news? Can algorithmic thinking, i.e. the way algorithms act as a result of their training, be transmitted to human beings and lead them to act in the same way?

To understand this notion, which will enable us to better analyze the impact of algorithmic thinking on the behaviors and strategies adopted by influencers, we need to answer this question: how do algorithms approach information processing and decision-making, in other words, what are the principles of their "algorithmic thinking"?

Secondly, we need to ask whether, following prolonged exposure to algorithms, influencers begin to internalize algorithmic logic, to the point of thinking and acting in accordance with it.

Influenced by algorithmic logic, do influencers conform to the preferences dictated by the algorithms, which could lead to censorship and self-censorship phenomena and, consequently, a depletion of available data?

What are the effects of selection and display algorithms on the diversity and quality of the content presented in our news feeds?

8.3. Theoretical points of reference

The result of a "set of techniques enabling machines to perform tasks and solve problems normally reserved for human beings" (Le Cun 2016, author's translation) thanks to their machine learning capacity, AI algorithms have been supporting users on social networks since 2010, selecting and displaying the content each user consumes on their account, intermediating their exchanges and directing them towards advertising. Their aim is to satisfy users' expectations, even before they are expressed. To achieve this, AI algorithms use active observation of all user gestures, digital surveillance.

The introduction of algorithms into users' phones occurred when they created their first social networking account, without warning or instructions. Platforms (Facebook, Instagram, TikTok, etc.) invited human beings to open network accounts, promising that through this tool they would create links with other human beings. Initially, communication links were created exclusively between human beings (the main feature of Web 2.0). However, as we entered the period of Web 3.0, where intelligent devices began to replace human partners, platforms began to insert a third interlocutor into the pattern of human-to-human communication: the AI algorithm.

These AI algorithms began to develop little by little, thanks to each learning experiment carried out by observing the actions of networked human beings (conversations and actions). Today, the algorithms launched in 2010 (those still "alive") have already learned enough to become the third active interlocutor, following Paul Lazarsfeld's (1901–1976) scheme, with the ability to select and even interpret the information communicated by the sender (before communicating it to others). "Algorithmic recommendation systems enable platforms to regulate user content by promoting and amplifying certain types of content considered more relevant and appropriate" (Zeng et al. 2022, author's translation). It is the algorithm that decides what users will see on the walls of their accounts, the order of information that will be displayed to them – gestures similar to the actions of TV news editors who select and organize the information presented to viewers – the user wall on social networks being the equivalent of a major TV newscast, having the same effect, the agenda effect: the construction of a worldview, the construction of reality.

Algorithms, like journalists, become Pierre Bourdieu's (1993–2003) tinkerers with reality in an environment marked by device dependency. A reality that, in a type of immersive consumption, attracts a younger audience than in the traditional press (TV, print, etc.). The annual Kantar Public Onepoint barometer for *La Croix* shows that the under-35s get their information primarily from the Internet (66%), mainly via smartphones, followed by television (only 26%). The written press is very much in the minority (1%). Among the under-25s, according to the survey conducted by *Étude Diplomeo* in 2020, out of 4,682 young people aged 16 to 25, 73% use the Internet for information, including 36% on *Instagram*, 29% on *Twitter* and 28% on Facebook. Since 2020, during the Covid-19 pandemic, users have migrated from Instagram to TikTok, where they encounter the very powerful AI algorithms, the most powerful on the market.

Declared the most addictive "due to its extremely advanced and sophisticated algorithmic system" (Smith and Short 2022, author's translation), TikTok, a video-sharing platform launched in 2016 with over 2.6 billion downloads, "allowing users to share short video clips ranging from two seconds to three minutes" (Chen et al. 2021; Bhandari and Bimo 2022), is also "the social network most used by Generation Z worldwide, with over one billion monthly users" (Boeker and Urman 2022; Leung et al. 2022). In January, there were 21 million TikTok users in France[1].

TikTok's algorithms are not neutral: a whistleblower has revealed to the *Wall Street Journal* that TikTok's algorithm demonstrates significant manipulation and restriction in determining the content presented to users, with the aim of prolonging their engagement on the platform. Leaked internal documents show that "TikTok takes international measures to limit

1 Available at: www.statista.com/statistics [Accessed 26 October 2023].

the visibility of videos featuring individuals with features considered 'unattractive' in appearance, age or body shape, so as not to be seen outside their country of origin" (Zeng et al. 2022, author's translation). Douyin also encourages creators to specialize in a single field, rather than seeking to cover a wide range of content. TikTok's algorithm has no direct knowledge of a user's personal values but makes assumptions based on user behavior and interactions. It does not just monotonously recommend what users like.

Algorithm favoritism in the "for you" algorithm can introduce its own biases and values into the design, "by which users can be unknowingly influenced" (Harriger et al. 2022, author's translation). Algorithmic filtering in personalized technology has long been debated for its potential to limit users' exposure to diverse information.

8.3.1. *Algorithmic thinking*

In line with the theory of the Extended Mind (Clark and Chalmers 1998), "the environment plays an active role in how cognitive processes are directed" (author's translation). In this respect, non-biological systems such as algorithms can be seen as part of a cognitive process in which our brains perform certain operations, while "the algorithm also plays a causal role in the governance of an individual's personal values and behaviors, as an interactive system or cognitive integration" (Menary 2010, author's translation; Chakravarty 2021).

"Algorithmic thinking" can describe the underlying logic that guides the operation of algorithms. Although AI algorithms do not "think" in the human sense, they are designed to solve problems or make decisions according to certain mathematical models or rules.

8.3.2. *"Pleasing the algorithms"*

Becoming an influencer means pleasing the algorithms. To achieve influencer status, we need to be favored by algorithms, the new regulators of the digital public space, which select and display content in users' accounts and monetize it.

The AI effect: as influencers' content is highlighted by the algorithm, it will become more attractive to advertisers, which can encourage influencers to optimize their publications according to algorithmic criteria rather than creative or informative merit. Money is a powerful motivator that can encourage influencers to create content that is popular but not necessarily aligned with their interests or talents.

Some of these areas may be more monetizable than others, which could encourage TikTok to highlight them to attract more advertising or commercial partnerships. However, this may also limit influencers' ability to diversify their brand or explore new areas.

8.4. Methodology

8.4.1. *First step: creating an AI algorithm*

Douyin has revealed the logic of its recommendation algorithm to the public and has even published official reports to guide creators. It wants creators to know what constitutes content preferred by the algorithms. This also helped me in the first stage of creating an algorithmic model: we created an AI algorithm simulator. To simulate TikTok's content recommendation algorithm, we first defined the objective of reproducing the platform's recommendation system. Next, we collected and prepared the relevant data, such as video metadata and user interactions. The AI model was developed and trained using Python with libraries such as Scikit-learn and TensorFlow, integrating content analysis and user preference modeling techniques. A user-friendly interface was designed with Flask to facilitate interactions, and rigorous testing was carried out to validate the simulator's performance. Finally, we documented and deployed the simulator for effective use in our research on TikTok algorithms.

The results answered the question: how do algorithms approach information processing and decision-making, in other words, what are the principles of their "algorithmic thinking"? Here they are, as follows:

– Heuristics-quantitative thinking (data-driven decision-making): the principle of reinforcement. When a user publishes content that receives positive attention (likes, shares, comments), this acts as a reward.

– Heuristic-optimization/single versus multiple objectives: algorithms often seek to optimize a certain objective function. This focus on certain objectives limits the algorithm's ability to diversify the content it presents to users.

– Heuristics-data: machine learning algorithms train on historical data. If TikTok's early users were particularly interested in these areas, the algorithm would have learned to prioritize this type of content.

– Exploration versus exploitation: the discovery of new domains would require an algorithm capable of processing and analyzing a greater variety of data, which can be costly in terms of computing time and resources.

– Temporality and immediacy: capturing and holding attention in the immediate. This can influence people to value immediacy and reactivity, rather than reflection.

– Heuristics-collaborative AI filtering: algorithms will look at what others are doing and propose the same content, which will not encourage diversity.

Following prolonged exposure to algorithms, do influencers begin to internalize algorithmic logic, to the point of thinking and acting in accordance with it?

8.4.2. *Second step: node analysis*

Node analysis was used to detect and study the actions of influencers in correlation with those of the algorithms. Network analysis is a methodological approach that explores the relationships between entities (called "nodes") and how these relationships (called "links" or "edges") structure the overall network.

– Nodes: accounts of individual influencers or groups of influencers.

– Links: types of relationships, interactions (likes, comments, retweets), mentions, mutual follow-ups, etc.

– Data collection: API for TikTok to collect data on interactions and relationships between influencers.

– Network analysis: NodeXL provides an overview of nodes and links.

– Centrality measures: identify the most influential or central nodes in the network, as well as influencers who dominate the network and may have internalized the algorithmic logic.

– Community detection: identify clusters or groups of influencers who frequently interact with each other. These communities can reveal groups of influencers who adopt similar behaviors.

– Network patterns: look for recurring patterns that could indicate algorithmic behavior. For example, if certain influencers always interact with a certain type of content or a certain group of influencers.

From August 15 to October 5, we tracked 112 accounts identified by the node method, starting with seven influencer accounts, and collected 120 messages.

8.4.3. *Third step: identify semantic universes on TikTok*

We used the TikTok development API to collect and analyze influencer publications, videos, tweets, etc., in order to detect recurring patterns or themes.

The same method was used as in our previous study of Facebook and Instagram (Badau 2018, p. 659): our study began with the initial codification of information and ended with theory validation procedures. The first stage involved translating some of the analyzed messages into narratives, which were then codified into specific terms.

In 2018, a search was conducted on Facebook (source) to discover the 50 semantic universes (the 50 universes of interest), and the results showed that users live in these universes: friends, discovery, love, hello, make-up, cat, dog, chocolate, food, drink, sex, thank you, beauty, health, fashion, danger, money, illness, baby, win, happiness, scandal, drama, sports, violence, justice, nudity, politics, family, free, travel, war, man/woman, chat, weather, music, nature, relationships, film, solidarity, gift and rich.

One year on, we followed the migration process initiated by users who left Facebook for Instagram to see if they took their semantic universes from Facebook onto Instagram. And we discovered that on Facebook there are only 27 semantic universes, an indicator of the impoverishment following the switch from Facebook to Instagram.

On Instagram, there are 27 semantic universes: make-up, presentation of a body part (K. Jenner), brand presentation, posing to be photographed (#posing), nature (vegetarian), money (wealth/extravagance), fashion, food (luxury or appetizing), car (luxury), family, dog, cat (#adorable), sport (fitness), music, entertainment, peace, visual art (#realpainting), creativity, architecture, handmade, adrenaline, humor, friendship and love/seduction. The Instagram platform offers plenty of room for individualism!

We continued this study as part of this research on TikTok to analyze the effects of AI algorithms on the semantic universes of this platform's users.

8.4.4. *Fourth step: identify existing "rabbit holes" in TikTok's semantic universes*

Douyin's collaborative filtering algorithm may inadvertently help users discover their latent interests, but this can also become a trap, with users finding themselves drawn into less regulated "rabbit holes" (Riemer and Peter 2021; Harriger et al. 2022), as the algorithm exploits their psychological needs.

Findings from a study in Sweden[2] reveal that users' algorithm-influenced motivations include: entertainment, information seeking, hobbies, social status, finding inspiration for hobbies, habitual use and discovering new hobbies.

A considerable portion of respondents to the questionnaire acknowledged their attempts to actively avoid certain types of content on TikTok but without much success. Despite their efforts to avoid certain types of content, they are not always protected from personalized content suggested by the algorithm, as Harriger et al. (2022) also point out.

8.5. Conclusion

8.5.1. *Internalizing algorithmic logic*

When exposed to these algorithms on a recurring basis, influencers can internalize the logic and decision-making methods of the algorithms, thus influencing their own thinking and decision-making. This internalization can be seen as a manifestation of "algorithmic thinking", where individuals adapt their behavior according to what they perceive to be the algorithm's preferences.

These traces of algorithmic thinking have been found in influencers:

– Like algorithm optimization, influencers will seek to "optimize" (modify) their behavior to obtain favorable results from an algorithm, to

2 Available at: www.diva-portal.org/smash/record.jsf?pid=diva2%3A1777846&dswid=8571.

maximize certain outcomes (e.g. getting more likes or shares). The influencer will be rewarded (just like the algorithms!) for conforming to a certain type or style of content.

– Heuristic-optimization on a specific objective: influencers will create and share content specifically designed to maximize engagement, even though this means sacrificing quality and diversity. The hunt for "likes" is tantamount to AI optimization.

– Quantitative thinking (data-driven decision-making): they will opt for content that has already received a large number of "likes" or shares, assuming that popularity is an indicator of quality or relevance (algorithmic thinking). They see the metrics of AI algorithms as measures of their own value or success.

– Heuristics and collaborative filtering: just as algorithms use collaborative filtering to compare user profiles and recommend content similar to what other "similar" users have liked, influencers observe what type of content is frequently promoted by the algorithm. They then imitate the trends and copy popular formats from other creators that have generated a large response, instead of creating their own content.

– Heuristic approach: choosing to read only articles with a certain number of views or comments means adopting simple rules to make decisions, simulating the way an algorithm works. This creates a kind of platform culture (impoverishing domains), which leads to the creation of a single common meaning (Badau 2019a, 2019b, p. 48).

– Virtuous (or vicious, depending on your point of view) circle: once a certain type of content becomes popular, it can create a positive feedback cycle where more creators produce this type of content and more users consume it, thus prompting the algorithm to promote it further. As a result of this feedback loop, once a domain becomes popular, it attracts more creators, creating a relatively small virtuous (or vicious, depending on your point of view) circle where popularity automatically reinforces algorithmic visibility, which can lead to the impoverishment of domains.

8.5.2. *The effect of algorithmic thinking on the semantic universes of TikTok users*

An even more pronounced impoverishment of semantic universes (users' areas of interest) can be seen: only 15 on TikTok, an accentuated impoverishment in comparison with the 27 semantic universes on Instagram and the 50 on Facebook. Here they are: humor, games, music, cinema, news, lifestyle, food, politics, popularized science, gaming, travel, sports, daily news and dance.

8.5.3. *The "rabbit holes" hidden by AI algorithms in semantic universes on TikTok*

Hidden in the semantic universes on TikTok were the less regulated "rabbit holes", i.e. the traps created by the algorithms, from the Swedish study (we have noted "RH" for rabbit holes and "TSU" for Tiktok semantic universes):

– Entertainment (TL): humor and games, music, cinema (TSU).

– Searching for information, entertainment, hobbies or social status (RH): news and lifestyle, politics and popularized science (TSU).

– Find inspiration for leisure (RH): food, lifestyle, games and gaming, travel and sports (TSU).

– Usual use (RH): daily news (TSU).

– Discovering new leisure activities (RH): gaming, games, food and lifestyle (TSU) as new interests to explore.

8.6. References

Badau, H.M. (2018). Les enjeux éthiques de la communication de l'information d'actualité sur les blogs et réseaux sociaux. *Revue française des sciences de l'information et de la communication*, 12.

Badau, H.M. (2019a). Intelligence artificielle et réseaux socionumériques : étude d'impact sur la dynamique des mouvements sociaux en France et en Roumanie. *Les Cahiers du Numérique*, 15, 211–226 [Online]. Available at: https://www.cairn.info/revue--2019-3-page-211.htm.

Badau, H.M. (2019b). L'apprentissage non-supervisé des algorithmes dotés avec Intelligence Artificielle limite ou encourage la créativité ? *Argumentum. Journal of the Seminar of Discursive Logic, Argumentation Theory and Rhetoric*, 17(2), 112–127.

Bessière, P. and Dousson, C. (2018). *Intelligence artificielle*. CNRS Éditions, Paris.

Bhandari, A. and Bimo, S. (2022). Why's everyone on TikTok now? The algorithmized self and the future of self-making on social media. *Social Media + Society*, 8(1). doi: 10.1177/20563051221086241.

Boeker, M. and Urman, A. (2022). An empirical investigation of personalization factors on TikTok. In *WWW '22: Proceedings of The ACM Web Conference 2022*. Association for Computing Machinery, New York, 2298–2309.

Chakravarty, P. (2021). The mind leak: An analysis of Clark & Chalmers extended mind hypothesis, 28 February. doi: 10.2139/ssrn.3794727.

Chen, W., Liu, H., Zhang, Y. (2021) AI-based tracking systems: Enhancing efficiency and accountability. *Journal of Business Analytics*, 4, 89–102.

Clark, A. and Chalmers, D. (1998). The extended mind. *Analysis*, 58(1), 7–19.

Goodfellow, I., Bengio, Y., Courville, A. (2016). *A Deep Learning*. MIT Press, Cambridge.

Harriger, J.A., Evans, J.A., Thompson, J.K., Tylka, T.L. (2022). The dangers of the rabbit hole: Reflections on social media as a portal into a distorted world of edited bodies and eating disorder risk and the role of algorithms. *Body Image*, 41.

Ilut, P. (1997). *Approche qualitative en SHS : concepts et méthodes*. Polirom, Iasi.

Le Cun, Y. (2016). Présentation au College de France, 4 February, Paris [Online]. Available at: https://www.college-de-france.fr/agenda/lecon-inaugurale/apprentissage-profond-une-revolution-en-intelligence-artificielle/apprentissage-profond-une-revolution-en-intelligence-artificielle [Accessed 24 June 2024].

Leung, F.F., Gu, F.F., Li, Y., Zhang, J.Z., Palmatier, R.W. (2022). Influencer marketing effectiveness. *Journal of Marketing*, 86(6), 93–115. doi: 10.1177/00222429221102889 [Original published 2022].

Malerba, F. (2020). *L'intelligence artificielle et ses implications éthiques*. PUF, Paris.

Menary, R. (2010). Introduction: The extended mind in focus. In *The Extended Mind*, Menary, R. (ed.). Boston Review, Cambridge, MA, 1–25. doi: 10.7551/mitpress/9780262014038.003.0001.

Mercier, A. (2018). Fake news et post-vérité : 20 textes pour comprendre la menace. *The Conversation*.

Müller, V.C. and Bostrom, N. (2016). Future progress in Artificial Intelligence: A survey of expert opinion. In *Fundamental Issues of Artificial Intelligence*, Müller, V.C. (ed.). Springer, New York, 553–570.

Pélissier, N., Domenget, J.-C., Coutant, A., Rasse, P. (2015). Questionner ensemble les usages et les dispositifs : pour une approche sociotechnique. In *IACR/AIERI World Congress*, Montreal.

Riemer, K. and Peter, S. (2021). Algorithmic audiencing: Why we need to rethink free speech on social media. *Journal of Information Technology*, 36(4), 4.

Russell, S. and Norvig, P. (2021). *Artificial Intelligence: A Modern Approach*. Pearson, London.

Smith, T. and Short, A. (2022). Needs affordance as a key factor in likelihood of problematic social media use: Validation, latent Profile analysis and comparison of TikTok and Facebook problematic use measures. *Addictive Behaviours*, 129, 107259. doi: 10.1016/j.addbeh.2022.107259.

Zeng, Z., Dai, H., Zhang, D.J., Zhang, H., Zhang, R., Xu, Z., Shen, Z-J.M. (2022). The impact of social nudges on user-generated content for social network platforms. *Management Science*, 69(9), 5189–5208.

9

The Cultural and Creative Industries in the Age of AI: Ethics and Responses from Multilateral Organizations

9.1. Artificial intelligence: a constantly changing environment

Artificial intelligence (AI) was designed to automate cognition. Today, it is widely used in the cultural milieu, influencing our choices, behavior and imagination (Manovic 2018). The last five years have seen a revolution in the use of AI. It is no longer employed exclusively in research fields such as mathematics, biology or genetics. It is also present in more visible areas of our daily lives such as commerce, the media or the Cultural and Creative Industries (CCI), as well as for professional and consumer applications and platforms in the creation of videos, images, text, sound, voice or can be found within social networking applications. Despite constant refinement in imitating human creativity, these algorithms "do not really innovate [...] do not invent, do not adopt behaviors they were not designed for" (Linden 2020, authors' translation), but the professional CCI community is concerned and warns about the threats of AI, including the illegal use of intellectual property.

In order to function, AI algorithms need to be fed by thousands of pieces of data, generally available on the Internet. The arrival of the first free and paid generative AI platforms has raised awareness of the vulnerability of sites hosting creative productions, the lack of professional ethics on the part of the companies behind these platforms and the non-existence of a legal

---

Chapter written by Vanessa LANDAVERDE KASTBERG and Matina MAGKOU.

framework to protect the work of CCI actors. In fact, this is one of the main demands of the writers' strike in the USA in 2023: a framed use of AI-based software for screenwriting. Another reason for writers' anger is that this software is fed with data from their own creations. This is a very common situation in CCI circles. Paradoxically, other professionals in these industries are constantly asking for more powerful, more independent, more "creative" tools.

Within this framework, our contribution aims to describe the ethical challenges concerning the use of AI-based tools in the CCI and how they are currently being addressed at the supranational level. In this study, we present an overview of the key ethical and societal issues of AI as they are currently debated within the United Nations Educational, Scientific and Cultural Organization (UNESCO) and the European Union (EU): what are the conceptual foundations and starting points of these two institutions regarding their commitment to ethical AI? How do these two institutions specifically position themselves with regard to AI and the CCI? What specific strategies are these institutions putting in place to reconcile the advancement of technology with the needs and rights of creators and professionals in the CCI? What is the role of CCI companies? What is the regulatory and punitive framework for companies abusing the use and exploitation of artists' data? Is there an ethical use model for these companies?

In order to understand the state of the art of the approaches put in place by the EU and UNESCO, we rely on a reading of documents and reports issued by the two supranational institutions on AI and more particularly on the ethical issues raised by its use. This documentary work is supplemented by interviews with representatives of these institutions, as well as with interest groups on CCIs, but for professional reasons, their names cannot be published. The analysis is completed by collective reflections from international symposia dealing with the future of AI in CCIs.

9.2. AI and the technologies used by CCIs: what are the effects?

AI encompasses a range of technological practices, such as Symbolic AI, which relies on human-created symbology such as language to create rules that can then be reproduced (translation or image generation), or Machine Learning (ML), which involves computational systems learning structures from data, with the aim of making predictions or decisions (ChatGPT). The combination of these two technological approaches: the growth of private and Open Source applications, and the increase in databases that are open (Wikimedia, Insee) or accessible with authorization or payment (*New York Times*), opens up an immense number of opportunities for the media and creative industries. This huge number of applications calls for a classification of uses, which is why in the white paper *AI in the Media and Creative Industries*, Caramiaux et al. propose an articulation of uses along three axes: creation, production and consumption/distribution. Creation consists of the construction of new media supports (sounds, visuals, etc.). Production refers to the editing of material. Consumption/broadcasting refers to the interaction between the distribution platform and the user, as well as the automation of content suggestions. Each technology has its own specific use, making it easy to collect the data needed to make it work and to analyze user interaction and usage in order to understand user behavior and offer multiple, innovative experiences. This constant evaluation of technologies makes these applications more powerful, and therefore addictive and data-hungry.

In all three levels of technology, AI is highly dependent on human creativity and interaction. For a long time, it was thought that machines could not imitate human creativity (Li et al. 2020). However, if we look at the evolution of the latest technologies and take up the definition of the word creativity, as proposed by the *Larousse* dictionary: "capacity, faculty of invention, imagination; creative power", we see that the current situation of AI technologies is approaching this stage of innovation. This is leading us towards a new societal, professional and ethical crisis linked to new technologies, as was the case with social media. On the one hand, we have a part of society that is ready to cede its image rights, to accept all conditions of use in order to have fun with applications and platforms that allow them to play with photos taken with their smartphones and to create content (text, image, sound, etc.). Also, professionals or artists publish their creative work on social media or websites for mediatization; this content can benefit from copyright, like journalistic work. On the other hand, the companies behind these platforms need data to train and improve their technologies, so that some companies proceed with dubious practices, constantly flagged up by journalists, politicians, non-governmental organizations (NGOs) and consumer protection associations. In some cases, companies have already been sued, as illustrated by the case of OpenAI, which was sued by titles such as the *Chicago Tribune* and the *New York Daily News* for using their articles to train algorithms without any authorization (*Washington Post*, April 2024). These new generations of algorithms need data from professional and even artistic content, so that they can imitate and produce professional-looking content as faithfully as possible, and thus make it pay. Such use, often without the respective permissions, violates intellectual property rights and copyrights. In view of these emerging issues, in 2021 UNESCO published a series of recommendations, particularly in relation to ethics and the measures that states need to adopt in order to create the conditions and a legislative framework to protect all stakeholders in the use of AI. At the EU level, these recommendations have been taken on board, and a regulatory framework came into existence in February 2024, with the result that companies such as Google and Meta are now being prosecuted on the old continent.

9.2.1. *Between human and technological creativity: new challenges, new ways of working*

Despite the many issues surrounding copyright infringement, many professionals agree that AI, and in particular generative AI, is contributing to an increase in creativity. In an interview published on the website of marketing magazine La Réclame to present a study on creatives and generative AI carried out by creative agency Wunderman Thompson, creative director Paul-Émile Raymond explains how, in architectural firms and communications agencies, these tools are already being used as supports and accelerators for inspiration by art directors and their collaborators. Academic literature suggests a paradigm shift (Zhou and Lee 2024) in the creative process that translates into a sharing of creative tasks between human and machine: in the brainstorming phase, these tools can develop "creativity exponentially" (Jean et al. 2024), generate value and then increase innovation potential at all levels (Nadeau and Jobin 2024). As this AI–human collaboration consolidates, the co-creation model, hitherto normally carried out only between humans, is beginning to be accepted and seen as an asset (Wu et al. 2021). In its current state, generative AI does not possess any real creative skills; it is limited to proposing improved content, based on the data that has fed it. Professions in the creative industries will have to go through a period of adaptation, if they still have a future, as some are already advocating that creative work carried out exclusively by humans should be regarded almost as a luxury product. In this context, to encourage creative industry professionals not to give up their activity, companies and government institutions need to propose an ethical framework of protection for production and intellectual property.

9.2.2. *A question of ethics*

The new role of AI in the creative process, particularly in automation, sketch generation and experimentation, is indisputable, but the place of the artist remains crucial (Kulesz and Dutoit 2019). Experiments along these lines are already underway, as demonstrated by the work of art collective and research team in art and AI, *Obvious* (Sorbonne University). Since 2023, team members have been conducting experiments in innovative art production with the help of generative AI. Within a very precise framework, the team delimits human realization and realization by AI, making this experiment an interesting example of the ethical use of this technology in the creative industries. In order to understand what is at stake when working with these technologies, a non-exhaustive list of identified ethical issues is presented here (Huang et al. 2022):

– Transparency: the author and the creative process are identified.

– Anonymity: the author may or may not wish to be identified.

– Intention: the work created with AI is used for the purpose for which it was conceived.

– Responsibility: the author and stakeholders are identified and are prepared to take their share of responsibility in the event of a problem.

– Data origin: data sources are clearly identified.

– Understanding of algorithms: the author has prepared an accessible guide explaining how they work.

– Data protection: the author undertakes not to share the data used in their algorithm or that of users.

– Justice: authors of works that have fed the algorithms are recognized and share in the profits.

– Respect for intellectual property;

– Respect for copyright.

As we have seen, AI is challenging the foundations of the value chain within the CCI. In general, the ethical challenges posed by the use of AI and CCIs can be summarized as follows:

– Firstly, artists are increasingly concerned about the use of their largely uncredited and uncompensated work to shape these technologies. They also fear that these technologies herald the end of many creative jobs.

– Secondly, the integration of AI into CCIs must avoid cultural standardization and be geared towards promoting the diversity and variety of cultural expressions, which requires a clear ethical framework for algorithms to protect cultural and linguistic diversity.

The concerns expressed by CCI have inspired governmental and supra-governmental institutions at the European level to reflect on a legal framework that protects the different levels of users (consumer, producer, creator, broadcaster, etc.). At this level, it is important to mention UNESCO's important contribution to the debate on ideas and the protection of CCIs. After several years of discussion, in May 2024, the new law on AI was adopted by the EU member states, a first for the world and one that could inspire other states. In the following section, we present the background to this law and the criteria taken into account.

9.3. UNESCO, IA and CCI

In 2019, UNESCO's General Conference adopted 40 C/Resolution 37, mandating the organization to draw up an international standard-setting instrument on the ethics of AI in the form of a recommendation. The starting point for UNESCO to embark on this process was the conviction that AI affects all UNESCO's fields of competence, so that UNESCO finds itself at the heart of the transformations linked to the development of AI, according to a humanistic approach based on respect for human rights. The resolution was based on a preliminary study on the ethics of AI carried out by UNESCO's World Commission on the Ethics of Scientific Knowledge and Technology (COMEST) in the same year. The study had examined the drawbacks of AI technologies, in particular the risk of advancing business models that contribute to increasing already high levels of inequality within and between countries and that of further reducing diversity (including cultural diversity). Among the most crucial AI-related issues identified in this report is the standardization of cultural production.

Subsequently, in November 2021, UNESCO adopted its "Recommendation on the Ethics of AI". It calls for AI to be responsible, transparent and compliant with the rule of law. UNESCO's starting point was that "risks and ethical concerns should not hamper innovation and development but rather provide new opportunities and stimulate ethically conducted research and innovation that anchor AI technologies in human rights and fundamental freedoms, values and principles, and moral and ethical reflection" (UNESCO 2021).

UNESCO has also justified its involvement in this field by referring to the international normative framework, from the Universal Declaration of Human Rights (1948) to the Convention on the Protection and Promotion of the Diversity of Cultural Expressions (2005). Through its international role, UNESCO has also emphasized the need to pay particular attention to low- and middle-income countries, as "they have their own capacity but have been underrepresented in the AI ethics debate, which raises concerns about neglecting local knowledge, cultural pluralism, value systems and the demands of global fairness to deal with the positive and negative impacts of AI technologies" (UNESCO 2021).

The recommendation touches on all of UNESCO's core areas and the wider ethical implications of AI systems. With regard to culture, UNESCO stresses that AI technologies can enrich CCI, and also "Cultural identity and diversity, as AI technologies can enrich CCI, but can also lead to an increased concentration of supply of cultural content, data, markets and income in the hands of only a few actors, with potential negative implications for the diversity and pluralism of languages, media, cultural expressions, participation and equality" (UNESCO 2021). Culture is therefore one of the strategic areas identified in this recommendation. Although in the preamble to the recommendation, UNESCO refers to CCI, in the text concerning culture, identified as Strategic Area 7, there is no reference to CCI. Instead, UNESCO calls on member states, to examine and address the impact of AI systems on culture, particularly those of natural language processing applications, such as machine translation and voice assistants. The focus is on cultural diversity, an area in which UNESCO is positioned, and highlights the challenges of AI use for the disappearance of endangered languages, tonal and cultural variations associated with human expression and language. Two other important points raised by UNESCO are the call for member states to promote AI education, aimed at artists and creative professionals, so that they are able to assess the relevance of using AI technologies in their professions.

To follow up on these developments at the normative level, UNESCO organized a first high-level meeting on AI in the CCI , in Paris on November 19, 2023. The focus of the meeting was the impact of AI in the audiovisual sector. The strikes by American screenwriters and actors that have been going on for several months were the main reason for the meeting. It also comes at a very important time, as the United States of America has rejoined the organization in July 2023. In the letter sent to UNESCO's Director-General at the beginning of June 2023, the United States refers to the reforms and initiatives that motivated its return, including UNESCO's new initiative on the ethics of AI. All these initiatives by the organization have begun to yield concrete results, notably at the level of the EU.

9.3.1. *EU, AI and CCI*

On March 9, 2018, the European Commission issued a press release entitled "Artificial Intelligence", whereby the Commission kick started work to align cutting-edge technology with ethical standards (Commission européenne 2018). The statement announced that the Commission was setting up a group to work on AI by gathering expert opinions and rallying a broad alliance of stakeholders. The press release marked a first step in the debate on the governance of AI. In April 2018, the Commission unveiled not only a new European AI strategy but also an action plan for AI governance, and the creation of a High-Level Group on AI to make recommendations on ethical guidelines for AI-related challenges.

This elite group's recommendations would then feed into the policy-making process and be used to refine the EU's approach to AI governance, with the creation of a "white paper" on AI as well as a coordinated AI plan in April 2021. The latter led to the creation of a new AI law: a European law that would subject AI to specific legal requirements based on market economy principles and risk assessment.

In 2022, the European Commission's Directorate-General for Education and Culture and Directorate-General Connect commissioned a study entitled "Opportunities and Challenges of Artificial Intelligence Technologies for the Cultural and Creative Sectors". The study stresses that despite the fact that AI has the potential to transform the creative sector, it is essential to weigh up the advantages and disadvantages of its use. The challenges identified are grouped approximately five areas, including access to data; access to skills; transparency; collaborative ecosystems; and access to finance, taking into account the wide diversity of needs across CCIs, and particularly those of smaller actors. The study also highlights the importance of data interoperability.

After months of extensive discussions and consultations, reading and drafting reports and studies, in May 2024, EU policymakers were able to propose a coherent legislative framework capable of protecting the various users and framing the companies and institutions behind the development of these systems. The AI Act, the world's first legislation to regulate AI, came into being following a negotiation process by the "trialogue", i.e. Europe's three key institutions – the European Parliament, the European Council and the European Commission. These three institutions worked out the details of the new laws while being assailed from all sides by lobbyists, including those from the CCI, and also trade unions hoping to influence the final legislation.

The interviews conducted show that representatives of workers' interest groups in the various CCI sectors at the European level consider that the AI law will fail to meet their needs and that the protection of artists and performers will remain at the heart of the sector's demands in the years to come. They stress that there must be explicit provision for appropriate transparency requirements in AI law, including robust transparency obligations regarding the use of copyrighted content by generative AI models.

For the EU, in order for AI law to accomplish its missions, it needs to have strong European tools and actors in this field. Also, the ethical framework and guidelines must be compatible with EU principles and regulatory frameworks. In this respect, it is worth recalling that the agreement recently reached between the European institutions opens up a new phase of regulatory tension and resistance in the European market. To ease this situation, Luciano Floridi, Professor of Philosophy of Ethics and Information at Oxford University, has proposed that the EU use "soft" digital ethics to support regulation and legislation and balance different values and interests (Floridi 2019).

9.3.2. *Protecting creation and creativity: a question of ethics*

UNESCO's recommendations on AI and the necessary legal framework have been adopted within the EU. These efforts attest to the importance of multilateralism in the age of technology (Pauwels 2019), and it remains to be seen whether other countries will follow suit. AI knows no borders or application limits in its possibilities and has an impact on the international exchange of digital cultural goods and services. All these initiatives are major developments, as they initiate discussions on the ethics of AI at both national and international levels. They will enable the application of their results in socio-political strategies, for the application of AI's ethical and social values (Cath et al. 2018).

9.3.3. *New business models in CCI through the sensible use of AI*

CCI companies did not wait for the arrival of a legislative framework to self-regulate and apply ethical practices to the way they operate. Although institutional efforts have contributed to this awareness, the general public, professionals and creatives have pushed companies to review their practices. The rising negative image of certain companies through social media, press articles and negative opinions has influenced the advancement of ethical practices. Since 2023, new business models have begun to emerge, in an effort to reconcile creators, consumers and distributors. In order to avoid being perceived as data "thieves", these models emphasize their ability to reward the sources of their data: creators and websites. Such is the case with Shutterstock, the leading platform for royalty-free or paid image databases. In 2023, the company launched its TRUST (Training, Royalties, Uplift, Safeguards and Transparency) program. The site describes its program in these terms: training its algorithms with licensed data; royalties that fairly remunerate artists; enhancing and promoting images that represent and reflect the world; safeguards to control content and protect consumers, with transparency as the default in its practices. Therefore, the images on the site created from generative AI are derived from algorithms using data obtained legally and ethically. This makes the company one of the pioneers in the ethical use of AI.

9.4. What response can we give to CCI professionals on the future of ethics in creativity?

We have tried to present and explain the major issues at stake in the use of AI in CCIs, as well as the many efforts being made by international institutions up to spring 2024. The bulk of the new legislative framework will not come into force until 2026, as many societal, commercial and technological changes may occur during this period, and the law may become outdated before it is implemented. However, we have seen how artist collectives or companies have not waited for regulation to create their own ethical framework for AI use. This is perhaps the most realistic way forward at the moment, and while we wait for a global awakening on the part of states and other organizations working for the ethical use of AI. One thing is certain: without data, AI can neither advance nor make innovative proposals, whereas human creativity just needs imagination and inspiration for endless production. In this period of transition and paradigm shift, as technology continues to perfect its imitation of human creativity, it is the responsibility of CCI professionals to reflect on how to reinvent or readapt their creative skills for an age that wants to be automated.

9.5. References

Anantrasirichai, N. and Bull, D. (2022). Artificial intelligence in the creative industries: A review. *Artificial Intelligence Review*, 55, 589–656.

Caramiaux, B., Lotte, F., Geurts, J., Amato, G., Behrmann, M. Bimbot, F., Falchi, F., Garcia, A., Gibert, J., Gravier, G. et al. (2019). AI in the media and creative industries. Research Report, New European Media (NEM), hal-02125504.

Cath, C., Wachter, S., Mittelstadt, B., Taddeo, M., Floridi, L. (2018). Artificial Intelligence and the "Good Society": The US, EU, and UK approach. *Science and Engineering Ethics*, 24, 505–528. doi: 10.1007/s11948-017-9901-7.

Commission Européenne (2018). Intelligence artificielle – La Commission entame des travaux pour conjuguer technologies de pointe et principes éthiques. Press Release, 9 March [Online]. Available at: https://ec.europa.eu/commission/presscorner/detail/fr/IP_18_1381 [Accessed 13 July 2024].

Commission Européenne (2022). Opportunities and challenges of artificial intelligence technologies for the cultural and creative sectors. Publications Office of the European Union [Online]. Available at: https://digital-strategy.ec.europa.eu/en/library/opportunities-and-challenges-artificial-intelligence-technologies-cultural-and-creative-sectors [Accessed 11 July 2024].

Dessalles, J.-L. (2019). *Des intelligences TRÈS artificielles*. Odile Jacob, Paris.

Dictionnaire Larouse (2024). Définition du mot "créativité" [Online]. Available at: https://www.larousse.fr/dictionnaires/francais/cr%C3%A9ativit%C3%A9/20301 [Accessed 13 July 2024].

Floridi, L. (2019). Soft ethics and the governance of the digital. *Philosophy and Technology*, 31(1). doi: 10.1098/rsta.2018.0081.

Gefen, A. (ed.) (2023). *Créativités artificielles. La littérature et l'art à l'heure de l'intelligence artificielle*. Les presses du réel, Dijon.

Huang, C., Zhang, Z., Mao, B., Yao, X. (2022). An overview of artificial intelligence ethics. *IEEE Transactions on Artificial Intelligence*, 4(4), 1–21. doi: 10.1109/tai.2022.3194503.

Jean, A., Luc, L., de Brabandère, L., Saive, M., Doutrepont, T. (2024). 5 mythes à déconstruire sur l'IA générative et la créativité. *HBR France* [Online]. Available at: https://www.hbrfrance.fr/innovation/5-mythes-a-deconstruire-sur-lia-generative-et-la-creativite-60439 [Accessed 13 July 2024].

Kulesz, O. and Dutoit, T. (2019) L'intelligence artificielle dans l'art et les industries culturelles et créatives : panorama des technologies, expertises et bonnes pratiques dans l'espace francophone. Report [Online]. Available at: https://www.francophonieinnovation.org/data/sources/users/1465/20201022154901-oifguide-ia-artvf.pdf [Accessed 13 July 2024].

Li, Y., Gu, J., Wang, L. (2020). Research on artificial intelligence ethics in the field of art design. *Journal of Physics: Conference Series*, 1673(1), 12052. doi: 10.1088/1742-6596/1673/1/012052.

Linden, I. (2020). Entre rêves et illusions… L'intelligence artificielle en question. *Revue d'éthique et de théologie morale*, 307, 11–27. doi: 10.3917/retm.310.0011.

Manovich, L. (2018). *AI Aesthetics*. Strelka Press, Moscow.

Nadeau, P. and Jobin, K. (2024). Enjeux éthiques et sociétaux de l'IA. In *Intelligence artificielle : génération générative : ChatGPT, Midjourney... S'approprier les nouvelles IA qui révolutionnent le monde professionnel*, Nadeau, P. and Jobin, K. (eds). Dunod, Paris, 115–169.

Parlement européen (2024). Artificial intelligence act [Online]. Available at: https://www.europarl.europa.eu/doceo/document/TA-9-2024-0138_EN.pdf [Accessed 13 July 2024].

Pauwels, E. (2019). How can multilateralism survive the era of artificial intelligence. *UN Chronicle*, 55(4), 50–54 [Online]. Available at: https://www.un-ilibrary.org/content/journals/15643913/55/4/14 [Accessed 13 July 2024].

Shutterstock, Inc. (2023). Shutterstock Unveils TRUST, its Best-in-Class Approach for Ethical AI [Online]. Available at: https://investor.shutterstock.com/news-releases/news-release-details/shutterstock-unveils-trust-its-best-class-approach-ethical-ai [Accessed 11 July 2024].

Unesco (2021). Recommandation sur l'éthique de l'Intelligence Artificielle [Online]. Available at: https://www.unesco.org/fr/artificial-intelligence/recommendation-ethics [Accessed 13 July 2024].

Vynck, G.D. (2024). 8 major newspapers join legal backlash against OpenAI, Microsoft. *Washington Post*, 30 April [Online]. Available at: https://www.washingtonpost.com/technology/2024/04/30/chicago-tribune-open-ai-microsoft-lawsuit/ [Accessed 13 July 2024].

World Economic Forum (2024). Arts and culture: Cultural and creative industries [Online]. Available at: https://intelligence.weforum.org/topics/a1Gb0000002V7TyEAK/key-issues/a1Gb0000002V7UNEA0 [Accessed 13 July 2024].

Wu, Z., Ji, D., Yu, K., Zeng, X., Wu, D., Shidujaman, M. (2021). AI creativity and the human-ai co-creation model. In *Human-Computer Interaction. Theory, Methods and Tools*, Kurosu, M. (ed.). Springer, Cham, 171–190. doi: 10.1007/978-3-030-78462-1_13.

Zewe, A. (2023). Explained: Generative AI. *MIT News* [Online]. Available at: https://news.mit.edu/2023/explained-generative-ai-1109 [Accessed 13 July 2024].

Zhou, E. and Lee, D. (2024). Generative artificial intelligence, human creativity, and art. *PNAS Nexus*, 3(3), 52–60.

List of Authors

Horea Mihai BADAU
CREM
Université de Lorraine
Metz
France

André BOYER
GRM
Université Côte d'Azur
Nice
France

Alexandre CARBONNEAU
SIC.Lab Méditerranée
Université Côte d'Azur
Nice
France

Juan Antonio CARRASCO
Universitat Autònoma de Barcelona
Spain

Franck DEBOS
SIC.Lab Méditerranée
Université Côte d'Azur
Nice
France

Faranak FARZANEH
IPAG
Nice
France

Maria Daniella LANCINI
SIC.Lab Méditerranée
Université Côte d'Azur
Nice
France

Vanessa LANDAVERDE KASTBERG
SIC.Lab Méditerranée
Université Côte d'Azur
Nice
France

Matina MAGKOU
SIC.Lab Méditerranée
Université Côte d'Azur
Nice
France

Nicolas OLIVERI
SIC.Lab Méditerranée
Université Côte d'Azur
Nice
France

Loïc SUMIEN
SIC.Lab Méditerranée
Université Côte d'Azur
Nice
France

Index

E, F

G, H

I, L

M, N

O, P, R

S, T

U, V

Other titles from

iSTE

in

Science, Society and New Technologies

2025

ACCART Jean-Philippe
Marketing for Libraries and Information Services: Systems and Developments

CHRISTOFLE Sylvie
Tourism Agility in Times of Crisis and Uncertainty 1: Between Technology and Society (Tourism and Mobility Systems Set – Volume 11)
Tourism Agility in Times of Crisis and Uncertainty 2: Between Marketing-Communication and Sustainability (Tourism and Mobility Systems Set – Volume 12)

LEVEAU Pierre-Henry
Tourism, Technologies and Consumption in the 5.0 Era: Discovery and Reflection in the Face of Organizational and Societal Issues (Tourism and Mobility Systems Set – Volume 10)

MEATCHI Sourou
Revenue Management in the Age of Artificial Intelligence: Towards Ethical and Responsible Price Management in Air Transport, Tourism and Hospitality (Tourism and Mobility Systems Set – Volume 13)

N'GOALA Gilles, PAULHIAC SCHERRER Florence, DURIF Fabien
The Smart and Resilient City: Models and Ways of Life

2024

BARTHES Angela, CIBIEN Catherine, ROMAGNY Bruno
Biosphere Reserves and Sustainable Development Goals 1: Scientific and Practical Educational Issues in the Mediterranean
(Territory Development Set – Volume 3)
Biosphere Reserves and Sustainable Development Goals 2: Issues, Tensions, Processes and Governance in the Mediterranean
(Territory Development Set – Volume 4)

BOTERO Nataly, LEDOUBLE Hélène, ALLARD-HUVER François
Pesticides: Pluridisciplinary Dialogues in Social and Human Sciences
(Communication, Environment, Science and Society Set – Volume 4)

CATELLANI Andrea, LITS Grégoire
Transitions in Tension: Controversies and Tensions Around Ecological Transitions
(Communication, Environment, Science and Society Set – Volume 3)

DAVALLON Jean
Heritage Traces in the Making: A Communicational Analysis of Modes of Heritagization
(Traces Set – Volume 6)

DEJEAN Maxime, L'HOSTIS Marine, LI Meng
Chinese Outbound Tourism: From Source Markets to Destinations
(Tourism and Mobility Systems Set – Volume 8)

GAY Jean-Christophe
Tourismophobia: From "Mass Tourism" to "Overtourism"
(Tourism and Mobility Systems Set – Volume 9)

JULIA Guilhem, FAUCHON Anne, KANAWATI Rushed
AI, Healthcare and Law

KOHLMANN Émilie
Linking with Nature in the Digital Age
(Communication, Environment, Science and Society Set – Volume 1)

LEDOUBLE Hélène
Popularizing Science: The Complex Terminological Interactions between Scientific and Press Discourses within the Field of Agroecology
(Communication, Environment, Science and Society Set – Volume 2)

LOUART Pierre, PIRIOU Jérôme, VIOLIER Philippe
Tourist Destinations According to Stakeholder Strategies: From Clusters to Local Tourist Systems
(Tourism and Mobility Systems Set – Volume 7)

2023

APCHAIN Thomas
Favela Tours: Building Otherness in the Favelas of Rio de Janeiro
(Tourism and Mobility Systems Set – Volume 6)

CARDON Alain, EL HAMI Abdelkhalak
Fundamental Generation Systems: Computer Science and Artificial Consciousness, the Informational Field of Generation of the Universe, the Sixth Sense of Living Beings
(Digital Sciences Set – Volume 4)

CURULLI Irene G., IKIZ KAYA Deniz, KHAEFI Arghavan
Heritage Education for Climate Action
(Research in Architectural Education Set – Volume 2)

ELAMÉ Esoh
The Sustainable City in Africa Facing the Challenge of Liquid Sanitation
(Territory Development Set – Volume 2)

JURCZENKO Emmanuel
Climate Investing: New Strategies and Implementation Challenges

LYUBAREVA Inna, WALDECK Roger
Diversity of Methodological Approaches in Social Sciences: Example of the Analysis of Media and Online Information
(Modeling Methodologies in Social Sciences Set – Volume 3)

PARET Dominique, CRÉGO Pierre, SOLÈRE Pauline
Smart Patches: Biosensors, Graphene and Intra-Body Communications

VENTRE Daniel, GUILLOT Philippe
Electronic Communication Interception Technologies and Issues of Power

VENTRE Daniel, LOISEAU Hugo
Cybercrime during the SARS-CoV-2 Pandemic (2019-2022): Evolutions, Adaptations, Consequences
(Cybersecurity Set – Volume 3)

2022

AIT HADDOU Hassan, TOUBANOS Dimitri, VILLIEN Philippe
Ecological Transition in Education and Research

CARDON Alain
Information Organization of The Universe and Living Things: Generation of Space, Quantum and Molecular Elements, Coactive Generation of Living Organisms and Multiagent Model
(Digital Sciences Set – Volume 3)

CAULI Marie, FAVIER Laurence, JEANNAS Jean-Yves
Digital Dictionary

DAVERNE-BAILLY Carole, WITTORSKI Richard
Research Methodology in Education and Training: Postures, Practices and Forms
(Education Set – Volume 12)

ELAMÉ Esoh
Sustainable Intercultural Urbanism at the Service of the African City of Tomorrow
(Territory Development Set – Volume 1)

FLEURET Sébastien
A Back and Forth Between Tourism and Health: From Medical Tourism to Global Health
(Tourism and Mobility Systems Set – Volume 5)

KAMPELIS Nikos, KOLOKOTSA Denia
Smart Zero-energy Buildings and Communities for Smart Grids
(Engineering, Energy and Architecture Set – Volume 9)

2021

BARDIOT Clarisse
Performing Arts and Digital Humanities: From Traces to Data
(Traces Set – Volume 5)

BENSRHAIR Abdelaziz, BAPIN Thierry
From AI to Autonomous and Connected Vehicles: Advanced Driver-Assistance Systems (ADAS)
(Digital Sciences Set – Volume 2)

BÉRANGER Jérôme
Societal Responsibility of Artificial Intelligence: Towards an Ethical and Eco-responsible AI
(Technological Prospects and Social Applications Set – Volume 4)

CORDELIER Benoit, GALIBERT Oliver
Digital Health Communications
(Technological Prospective and Social Applications Set – Volume 5)

DOUAY Nicolas, MINJA Michael
Urban Planning for Transitions

GALINON-MÉLÉNEC Béatrice
The Trace Odyssey 1: A Journey Beyond Appearances
(Traces Set – Volume 4)

HENRY Antoine
Platform and Collective Intelligence: Digital Ecosystem of Organizations

KOLOKOTSA Denia, KAMPELIS Nikos
Smart Buildings, Smart Communities and Demand Response
(Engineering, Energy and Architecture Set – Volume 8)

LE LAY Stéphane, SAVIGNAC Emmanuelle, LÉNEL Pierre, FRANCES Jean
The Gamification of Society
(Research, Innovative Theories and Methods in SSH Set – Volume 2)

RADI Bouchaïb, EL HAMI Abdelkhalak
Optimizations and Programming: Linear, Non-linear, Dynamic, Stochastic and Applications with Matlab
(Digital Sciences Set – Volume 1)

2020

ALAKTIF Jamila, CALLENS Stéphane
Migration and Climate Change: From the Emergence of Human Cultures to Contemporary Management in Organizations

BARNOUIN Jacques
The World's Construction Mechanism: Trajectories, Imbalances and the Future of Societies
(Interdisciplinarity Between Biological Sciences and Social Sciences Set – Volume 4)

ÇAĞLAR Nur, CURULLI Irene G., SIPAHIOĞLU Işıl Ruhi, MAVROMATIDIS Lazaros E.
Thresholds in Architectural Education
(Engineering, Energy and Architecture Set – Volume 7)

DUBOIS Michel J.F.
Humans in the Making: In the Beginning was Technique
(Social Interdisciplinarity Set – Volume 4)

ETCHEVERRIA Olivier
The Restaurant, A Geographical Approach: From Invention to Gourmet Tourist Destinations
(Tourism and Mobility Systems Set – Volume 3)

GREFE Gwenaëlle, PEYRAT-GUILLARD Dominique
Shapes of Tourism Employment: HRM in the Worlds of Hotels and Air Transport
(Tourism and Mobility Systems Set – Volume 4)

JEANNERET Yves
The Trace Factory
(Traces Set – Volume 3)

KATSAFADOS Petros, MAVROMATIDIS Elias, SPYROU Christos
Numerical Weather Prediction and Data Assimilation
(Engineering, Energy and Architecture Set – Volume 6)

MARTI Caroline
Cultural Mediations of Brands: Unadvertization and Quest for Authority
(Communication Approaches to Commercial Mediation Set – Volume 1)

MAVROMATIDIS Lazaros
Climatic Heterotopias as Spaces of Inclusion: Sew Up the Urban Fabric
(Research in Architectural Education Set – Volume 1)

MOURATIDOU Eleni
Re-presentation Policies of the Fashion Industry: Discourse, Apparatus and Power
(Communication Approaches to Commercial Mediation Set – Volume 2)

SCHMITT Daniel, THÉBAULT Marine, BURCZYKOWSKI Ludovic
Image Beyond the Screen: Projection Mapping

VIOLIER Philippe, with the collaboration of TAUNAY Benjamin
The Tourist Places of the World
(Tourism and Mobility Systems Set – Volume 2)

2019

BRIANÇON Muriel
The Meaning of Otherness in Education: Stakes, Forms, Process, Thoughts and Transfers
(Education Set – Volume 3)

DESCHAMPS Jacqueline
Mediation: A Concept for Information and Communication Sciences (Concepts to Conceive 21st Century Society Set – Volume 1)

DOUSSET Laurent, PARK Sejin, GUILLE-ESCURET Georges
Kinship, Ecology and History: Renewal of Conjectures (Interdisciplinarity Between Biological Sciences and Social Sciences Set – Volume 3)

DUPONT Olivier
Power: A Concept for Information and Communication Sciences (Concepts to Conceive 21st Century Society Set – Volume 2)

FERRARATO Coline
Prospective Philosophy of Software: A Simondonian Study (Technological Prospects and Social Applications Set – Volume 3)

GUAAYBESS Tourya
The Media in Arab Countries: From Development Theories to Cooperation Policies

HAGÈGE Hélène
Education for Responsibility (Education Set – Volume 4)

LARDELLIER Pascal
The Ritual Institution of Society (Traces Set – Volume 2)

LARROCHE Valérie
The Dispositif: A Concept for Information and Communication Sciences (Concepts to Conceive 21st Century Society Set – Volume 3)

LATERRASSE Jean
Transport and Town Planning: The City in Search of Sustainable Development

LELEU-MERVIEL Sylvie, SCHMITT Daniel, USEILLE Philippe
From UXD to LivXD: Living eXperience Design

LENOIR Virgil Cristian
Ethically Structured Processes
(Innovation and Responsibility Set – Volume 4)

LOPEZ Fanny, PELLEGRINO Margot, COUTARD Olivier
Local Energy Autonomy: Spaces, Scales, Politics
(Urban Engineering Set – Volume 1)

METZGER Jean-Paul
Discourse: A Concept for Information and Communication Sciences
(Concepts to Conceive 21st Century Society Set – Volume 4)

MICHA Irini, VAIOU Dina
Alternative Takes to the City
(Engineering, Energy and Architecture Set – Volume 5)

PÉLISSIER Chrysta
Learner Support in Online Learning Environments

PIETTE Albert
Theoretical Anthropology or How to Observe a Human Being
(Research, Innovative Theories and Methods in SSH Set – Volume 1)

PIRIOU Jérôme
The Tourist Region: A Co-Construction of Tourism Stakeholders
(Tourism and Mobility Systems Set – Volume 1)

PUMAIN Denise
Geographical Modeling: Cities and Territories
(Modeling Methodologies in Social Sciences Set – Volume 2)

WALDECK Roger
Methods and Interdisciplinarity
(Modeling Methodologies in Social Sciences Set – Volume 1)

2018

BARTHES Angela, CHAMPOLLION Pierre, ALPE Yves
Evolutions of the Complex Relationship Between Education and Territories
(Education Set – Volume 1)

BÉRANGER Jérôme
The Algorithmic Code of Ethics: Ethics at the Bedside of the Digital Revolution
(Technological Prospects and Social Applications Set – Volume 2)

DUGUÉ Bernard
Time, Emergences and Communications
(Engineering, Energy and Architecture Set – Volume 4)

GEORGANTOPOULOU Christina G., GEORGANTOPOULOS George A.
Fluid Mechanics in Channel, Pipe and Aerodynamic Design Geometries 1
(Engineering, Energy and Architecture Set – Volume 2)
Fluid Mechanics in Channel, Pipe and Aerodynamic Design Geometries 2
(Engineering, Energy and Architecture Set – Volume 3)

GUILLE-ESCURET Georges
Social Structures and Natural Systems: Is a Scientific Assemblage Workable?
(Social Interdisciplinarity Set – Volume 2)

LARINI Michel, BARTHES Angela
Quantitative and Statistical Data in Education: From Data Collection to Data Processing
(Education Set – Volume 2)

LELEU-MERVIEL Sylvie
Informational Tracking
(Traces Set – Volume 1)

SALGUES Bruno
Society 5.0: Industry of the Future, Technologies, Methods and Tools
(Technological Prospects and Social Applications Set – Volume 1)

TRESTINI Marc
Modeling of Next Generation Digital Learning Environments: Complex Systems Theory

2017

ANICHINI Giulia, CARRARO Flavia, GESLIN Philippe,
GUILLE-ESCURET Georges
Technicity vs Scientificity: Complementarities and Rivalries
(Interdisciplinarity Between Biological Sciences and Social Sciences Set – Volume 2)

DUGUÉ Bernard
Information and the World Stage: From Philosophy to Science, the World of Forms and Communications
(Engineering, Energy and Architecture Set – Volume 1)

GESLIN Philippe
Inside Anthropotechnology: User and Culture Centered Experience
(Social Interdisciplinarity Set – Volume 1)

GORIA Stéphane
Methods and Tools for Creative Competitive Intelligence

KEMBELLEC Gérald, BROUDOUX Evelyne
Reading and Writing Knowledge in Scientific Communities: Digital Humanities and Knowledge Construction

MAESSCHALCK Marc
Reflexive Governance for Research and Innovative Knowledge
(Responsible Research and Innovation Set - Volume 6)

PARK Sejin, GUILLE-ESCURET Georges
Sociobiology vs Socioecology: Consequences of an Unraveling Debate
(Interdisciplinarity Between Biological Sciences and Social Sciences Set – Volume 1)

PELLÉ Sophie
Business, Innovation and Responsibility
(Responsible Research and Innovation Set – Volume 7)

2016

ANDRÉ Michel, SAMARAS Zissis
Energy and Environment
(Research for Innovative Transports Set – Volume 1)

BLANQUART Corinne, CLAUSEN Uwe, JACOB Bernard
Towards Innovative Freight and Logistics
(Research for Innovative Transports Set – Volume 2)

COHEN Simon, YANNIS George
Traffic Management
(Research for Innovative Transports Set – Volume 3)

EL FALLAH SEGHROUCHNI Amal, ISHIKAWA Fuyuki, HÉRAULT Laurent, TOKUDA Hideyuki
Enablers for Smart Cities

GIANNI Robert
Responsibility and Freedom: The Ethical Realm of RRI
(Responsible Research and Innovation Set – Volume 2)

GRUNWALD Armin
The Hermeneutic Side of Responsible Research and Innovation
(Responsible Research and Innovation Set – Volume 5)

LENOIR Virgil Cristian
Ethical Efficiency: Responsibility and Contingency
(Responsible Research and Innovation Set – Volume 1)

PELLÉ Sophie, REBER Bernard
From Ethical Review to Responsible Research and Innovation
(Responsible Research and Innovation Set – Volume 3)

REBER Bernard
Precautionary Principle, Pluralism and Deliberation: Science and Ethics
(Responsible Research and Innovation Set – Volume 4)

TORRENTI Jean-Michel, LA TORRE Francesca
Materials and Infrastructures 1
(Research for Innovative Transports Set – Volume 5A
Materials and Infrastructures 2
(Research for Innovative Transports Set – Volume 5B)

VENTRE Daniel
Information Warfare – 2nd edition

YANNIS George, COHEN Simon
Traffic Safety
(Research for Innovative Transports Set – Volume 4)

2015

BRONNER Gérald
Belief and Misbelief Asymmetry on the Internet

2009

VENTRE Daniel
Information Warfare

Printed and bound by CPI Group (UK) Ltd, Croydon, CR0 4YY
13/09/2026
14961011-0002